Lecture Notes in Computer Science 11393

Commenced Publication in 1973
Founding and Former Series Editors:
Gerhard Goos, Juris Hartmanis, and Jan van Leeuwen

More information about this series at http://www.springer.com/series/7409

Lijun Chang · Junhao Gan · Xin Cao (Eds.)

Databases Theory and Applications

30th Australasian Database Conference, ADC 2019
Sydney, NSW, Australia, January 29 – February 1, 2019
Proceedings

 Springer

Editors
Lijun Chang
University of Sydney
Sydney, NSW, Australia

Junhao Gan
University of Melbourne
Parkville, VIC, Australia

Xin Cao
University of New South Wales
Sydney, NSW, Australia

ISSN 0302-9743 ISSN 1611-3349 (electronic)
Lecture Notes in Computer Science
ISBN 978-3-030-12078-8 ISBN 978-3-030-12079-5 (eBook)
https://doi.org/10.1007/978-3-030-12079-5

Library of Congress Control Number: 2018967952

LNCS Sublibrary: SL3 – Information Systems and Applications, incl. Internet/Web, and HCI

This Springer imprint is published by the registered company Springer Nature Switzerland AG
The registered company address is: Gewerbestrasse 11, 6330 Cham, Switzerland

Preface

It is our pleasure to present to you the proceedings of the 30th Australasian Database Conference (ADC2019), which took place in Sydney, Australia. The Australasian Database Conference is an annual international forum for sharing the latest research advancements and novel applications of database systems, data-driven applications, and data analytics between researchers and practitioners from around the globe, particularly Australia and New Zealand. The mission of ADC is to share novel research solutions to problems of today's information society that fulfil the needs of heterogeneous applications and environments and to identify new issues and directions for future research and development work. ADC seeks papers from academia and industry presenting research on all practical and theoretical aspects of advanced database theory and applications, as well as case studies and implementation experiences. All topics related to database are of interest and within the scope of the conference. ADC gives researchers and practitioners a unique opportunity to share their perspectives with others interested in the various aspects of database systems.

As in previous years, the ADC 2019 Program Committee accepted papers considered as being of ADC quality without setting any predefined quota. The conference received 16 submissions and accepted ten papers, including nine full research papers and one demo paper. Each paper was peer reviewed in full by at least three independent reviewers, and in some cases four referees produced independent reviews. A conscious decision was made to select the papers for which all reviews were positive and favorable. The Program Committee that selected the papers consists of 35 members from around the globe, including Australia, China, Finland, Japan, the UK, and the USA, who were thorough and dedicated to the reviewing process.

We would like to thank all our colleagues who served on the Program Committee or acted as external reviewers. We would also like to thank all the authors who submitted their papers and the attendees. This conference is held for you, and we hope that with these proceedings, you can have an overview of this vibrant research community and its activities. We encourage you to make submissions to the next ADC conference and contribute to this community.

January 2019

Lijun Chang
Junhao Gan
Xin Cao

General Chair's Welcome Message

On behalf of the organizers and Steering Committee for ADC 2019, I am honored to welcome you to the proceedings of the conference. The Australasian Database Conference has an extensive history; this was the 30th occurrence of the conference. In the past decade, ADC has been held on the Gold Coast (2018), Brisbane (2017), Sydney (2016), Melbourne (2015), Brisbane (2014), Adelaide (2013), Melbourne (2012), Perth (2011), Brisbane (2010), Wellington (2009), and Wollongong (2008). This year, ADC was run under the umbrella of the Australasian Computer Science Week, organized at Macquarie University in Sydney.

We are especially grateful to A/Prof. Len Hamey of Macquarie University for arranging things so smoothly. The technical program was arranged by Dr. Lijun Chang (University of Sydney) and Dr. Junhao Gan (University of Melbourne), who managed the review process by a panel of distinguished researchers from many countries, and then selected the papers from 16 submissions. The proceedings publication was arranged and supervised by Dr. Xin Cao (University of New South Wales). We are all the beneficiaries of their dedication.

As well as the conference, whose papers are found here, we held a co-located workshop aimed at PhD students and early-career researchers, with a range of outstanding speakers, especially a keynote from Prof. Beng Chin Ooi (NUS). This all shows the vibrancy of the database research community in Australia and New Zealand, and contributes to its continuation.

Alan Fekete

Organization

General Chair

Alan Fekete   The University of Sydney, Australia

Program Committee Co-chairs

Lijun Chang   The University of Sydney, Australia
Junhao Gan   The University of Melbourne, Australia

Publication Chair

Xin Cao   The University of New South Wales, Australia

Steering Committee

Rao Kotagiri   The University of Melbourne, Australia
Timos Sellis   RMIT University, Australia
Gill Dobbie   The University of Auckland, New Zealand
Alan Fekete   The University of Sydney, Australia
Xuemin Lin   The University of New South Wales, Australia
Yanchun Zhang   Victoria University, Australia
Xiaofang Zhou   The University of Queensland, Australia

Program Committee

Zhifeng Bao   RMIT University, Australia
Renata Borovica-Gajic   The University of Melbourne, Australia
Huiping Cao   New Mexico State University, USA
Xin Cao   The University of New South Wales, Australia
Muhammad Aamir Cheema   Monash University, Australia
Lisi Chen   The University of Wollongong, Australia
Farhana Murtaza   The University of Melbourne, Australia
  Choudhury
Gianluca Demartini   The University of Queensland, Australia
Janusz Getta   The University of Wollongong, Australia
Yusuke Gotoh   Okayama University, Japan
Michael E. Houle   National Institute of Informatics, Japan
Wen Hua   The University of Queensland, Australia
Guangyan Huang   Deakin University, Australia
Zi Huang   The University of Queensland, Australia
Jianxin Li   The University of Western Australia, Australia

Lei Li	The University of Queensland, Australia
Rong-Hua Li	Beijing Institute of Technology, China
Jixue Liu	The University of South Australia, Australia
Jiaheng Lu	The University of Helsinki, Finland
Parth Nagarkar	New Mexico State University, USA
Quoc Viet Hung Nguyen	Griffith University, Australia
Jianzhong Qi	The University of Melbourne, Australia
Lu Qin	The University of Technology, Sydney, Australia
Junhu Wang	Griffith University, Australia
Sheng Wang	RMIT University, Australia
Sibo Wang	The University of Queensland, Australia
Yajun Yang	Tianjin University, China
Hongzhi Yin	The University of Queensland, Australia
Weiren Yu	Aston University, UK
Wenjie Zhang	The University of New South Wales, Australia
Ying Zhang	The University of Technology, Sydney, Australia
James Xi Zheng	Macquarie University, Australia
Rui Zhou	Swinburne University of Technology, Australia
Yi Zhou	The University of Technology, Sydney, Australia
Yuanyuan Zhu	Wuhan University, China

Contents

Full Research Papers

Batch Processing of Shortest Path Queries in Road Networks

Mengxuan Zhang$^{(\boxtimes)}$, Lei Li, Wen Hua, and Xiaofang Zhou

School of Information Technology and Electrical Engineering,
The University of Queensland, Brisbane, Australia
mengxuan.zhang@uqconnect.edu.au, {l.li3,w.hua}@uq.edu.au,
zxf@itee.uq.edu.au

Abstract. Shortest path algorithm is a foundation to various location-based services (LBS) and has been extensively studied in the literature. However, server-side shortest path calculation faces a severe scalability issue when the business expands and a huge amount of requests are submitted to the server simultaneously. Although a straightforward solution widely-adopted in current industry is to deploy more processing resources, in this work, we aim to improve the efficiency algorithmically by answering queries in a batch and reusing shareable computations. In particular, we generalize the goal-directed A^* algorithm to correctly solve the batch processing problem with localized destinations. We further propose two decomposition algorithms to deal with scenarios where the destinations are sparse. Extensive evaluations on a real-world road network verify the superiority of our algorithm compared with state-of-the-art methods.

Keywords: Shortest path · Batch process · Road network

1 Introduction

As the rapid spread of GPS-enabled devices, we have witnessed a blooming of various location-based services (LBS) such as map applications, commercial navigation products, O2O taxi business, etc. One of the most crucial techniques behind them is the shortest path algorithm which, given two locations s and t in a road network $G = (V, E)$, returns a route from s to t with the minimum network distance.

The shortest path problem has been extensively studied in both academia and industry. Existing approaches can be categorized depending on whether they use an index or not. Index approaches [1–9] pre-compute and store various auxiliary information so that the shortest path can be retrieved quickly once required. However, these methods are usually space-consuming and hard to adapt to network dynamics since it incurs huge computational cost to update or rebuild the index accordingly. Non-index approaches [10–12], on the contrary, search in the road network directly which is more applicable in a dynamic environment

© Springer Nature Switzerland AG 2019
L. Chang et al. (Eds.): ADC 2019, LNCS 11393, pp. 3–16, 2019.
https://doi.org/10.1007/978-3-030-12079-5_1

although not as efficient as the index counterparts. Considering the rapid changes in a real-life road network, the non-indexing algorithms are more appealing to industry and commercial LBS.

Shortest path queries can be handled at either client-side or server-side. However, in some applications such as O2O taxi and ride sharing, server-side processing is inevitable since they require information from other shortest path requests (e.g., nearby taxis and customers). Scalability becomes a severe issue in this situation when a huge amount of requests are submitted to the server simultaneously (e.g., 100K requests per minute during the rush hour). Naturally, more servers can be deployed in order to cope with such a swarm of queries. But is there a resource-efficient way to address the scalability issue? That is, can we algorithmically reduce the total computational cost of simultaneous shortest path queries?

To this end, we propose a batch processing algorithm by reusing shareable shortest path computations. Intuitively, when queries are localized (i.e., origins and destinations are centralized in a small region such as airports, train stations, shopping malls, etc.), there could be a large proportion of common computations among these queries. In other words, it is possible to answer all these queries within a single network search. Therefore, we study the problem of 1-N (i.e., the same origin with different destinations) shortest path queries in this work. It is worth noting that the 1-N algorithm can be easily adjusted to address other problems such as M-1 (by applying 1-N algorithm on the reversed road network) and M-N (by combining 1-N and M-1 algorithms).

Inspired by the goal-directed A^* algorithm, we generalize it to the A^*-$1N$ algorithm such that it can find shortest paths to N different but localized destinations in one search. We further extend our approach to deal with widespread destinations. In particular, we propose two decomposition strategies, i.e., angle-based and distance-based, to split the N destinations into several clusters and answer them separately using the A^*-$1N$ algorithm. Our contributions in this work can be summarized as follows:

- We study a new problem of batch shortest path queries, which is quite important to many real-world location-based applications;
- We improve the efficiency and scalability algorithmically by reusing shareable computations among batch queries;
- We propose a generalized A^*-$1N$ algorithm and two decomposition methods to speed up computation under different scenarios;
- We evaluate the effectiveness and efficiency of our algorithms comprehensively using real data.

The rest of the paper is organized as follows. Section 2 discusses the related work. Section 3 presents our generalized A^*-$1N$ algorithm and two decomposition approaches with correctness and performance analysis. An empirical study is conducted and reported in Sect. 4, followed by a brief conclusion in Sect. 5.

2 Related Work

The basis of most existing approaches is the *Dijkstra's Algorithm* [10]. It can find not only the shortest path, but also the fastest path given a departure time [11, 13,14]. However, its search space is nearly a circle, which results in a waste of time to traverse the obviously useless nodes. The *Bi-Dijkstra* algorithm reduces the running time by searching from the source and the destination simultaneously. Although its search space becomes two smaller circles, their sum could still be big especially when the graph density is uneven. In order to further reduce the search space, some heuristic algorithms like A^* [12] and *ALT* [1] are proposed. Nevertheless, since they are goal-directed and require a fixed destination, they can only answer the query between a pair of nodes at a time. Even if there are several queries from the same node simultaneously, we still have to run these algorithms multiple times.

After that, various index structures are proposed to either reduce the search space or return the result directly. The first kind is the *pruning techniques*, such as *Reach* [15], *Geographic Container* [16], *Arc-Flag* [3] and *CH* [2]. They can prune the search space by adding different additional information through pre-computing all-pairs or partially all-pairs shortest paths (like shortcuts, grid indicators and so on). The more additional information is added, the faster the actual search is. The second group is called *Hop Labeling* [4], which includes *IS-Label* [17], *Hub-Labeling* [18], *Pruned Landmark Labeling* [5], *Hop-Doubling* [19] and *H2H-Index* [6]. They can answer the shortest distance query in $O(1)$ time by combining the shortest distance from the source node to a hop node, and from this hop node to the destination. However, it takes at least cubic time to build and does not support path retrieval. The last stream is based on database techniques [7,20,21]. It stores all-pair shortest path in database and builds indexes to retrieve the result. Obviously, it is the fastest in query answering, but it is also the most time- and space-consuming. Furthermore, since all these index approaches suffer from a large amount of preprocessing time, they are mostly designed for the static environment and hard to adapt to the dynamic one. As the traffic condition changes over time, we cannot afford to rebuild any of these indexes.

Finally, only a few works try to solve the *batch shortest path* problem. Reza [22] puts all the sources nodes in one big priority queue and run the *Dijkstra* directly. All the intermediate results are settled by taking the average from the sources. The algorithm terminates when all the destinations are visited. Apparently, they cannot get the accurate result. Mahmud [23] breaks the batch search into three parts: from the sources to an intermediate node, from this intermediate node to another intermediate node, and from the second intermediate node to the targets. Although they propose several approaches to cluster the sources and targets, they still cannot answer the query correctly.

3   Batch Shortest Path Algorithms

In this section, we describe our algorithms in details. Firstly, we generalize the pairwise A^* algorithm to A^*-1N algorithm and prove its correctness in Sect. 3.1. Then we discuss the representative node selection in Sect. 3.2. After that, we present two decomposition approaches to deal with the cases when the target set distributes in a relatively large area in Sect. 3.3. Finally we briefly discuss the batch query answering in Sect. 3.4. Some important notations are listed in Table 1.

Table 1. Important notations

Notation	Meaning
$G(V, E)$	Graph G, Node set $V = \{v_i\}$, Edge set $E \subset \{(v_i, v_j)\}$
$w(v_i, v_j)$	Weight of edge (v_i, v_j)
$l(v_i, v_k)$	Estimated cost between v_i and v_k
$q_{s,t}$	Path finding query from start node s and target node t
$Q_{s,T}$	Batch path finding query with start node s and target set T
$p_{s,t}$	Shortest/Fastest path from s to t
$d_{s,t}$	The cost of shortest/fastest path $P_{s,t}$
T	Target node set and $T \subset V$
$d(v_i, v_j)$	Cost from v_i to v_j and not necessarily the least cost
$d_{v_k}(v_i, v_j)$	Estimated cost from v_i to v_j and through v_k

3.1 A^*-1N Algorithm

Given a graph $G(V, E)$ with coordinate geometry and a shortest path query $q_{s,t}$, the classic pairwise A^* algorithm chooses the target t as representative node which guides the search towards it. H is a priority queue ordered by heuristic distance $d_{v_i}(s, t) = d(s, v_i) + l(v_i, t)$ in ascending order. When t is popped out from H, the shortest distance from s to t is found.

In order to generalize the classic A^* to answer the multiple targets query, we first prove we can choose any node to be the representative node r apart from t.

Lemma 1. *Node v_i popping out from H with no descending value $d_{v_i}(s, r) = d(s, v_i) + l(v_i, r)$.*

Proof. Suppose that the node μ_1 popped out after μ is with smaller estimated distance, that is $d_{\mu_1}(s, r) < d_\mu(s, r)$. There are two situations.

Case one, μ_1 is not the adjacent node of μ or μ_1 is the adjacent node of μ with $d(s, \mu) + w(\mu, \mu_1) \geq d(s, \mu_1)$. Since $d(s, \mu_1)$ won't be refreshed when μ popped out and μ is popped out before μ_1, it is sure that $d_\mu(s, v) \leq d_{\mu_1}(s, r)$.

Case two, μ_1 is the adjacent node of μ with $d(s, \mu) + w(\mu, \mu_1) < d(s, \mu_1)$. When μ popped out, $d_\mu(s, r) = d(s, \mu) + l(\mu, r)$ and $d(s, \mu_1)$ is updated with $d(s, \mu_1) = d(s, \mu) + w(\mu, \mu_1)$. Then $d_{\mu_1}(s, r) = d(s, \mu) + w(\mu, \mu_1) + l(\mu_1, r)$. According to triangle inequality, $d_\mu(s, r) \leq d_{\mu_1}(s, r)$.

These two cases are both contradictory with the assumption, then it is obvious that node v_i popped out from H with no descending value $d_{v_i}(s, r)$.

Theorem 1. *When node v_i popped out from H, $d(s, v_i) = d_{s, v_i}$.*

Proof. For any node v_i popped out with distance $d_{v_i}(s, r) = d(s, v_i) + l(v_i, r)$, there won't exits another smaller estimated value $d_{v_i}(s, r) = d(s, v_i) + l(v_i, r)$ according to Lemma 1. So when v_i popped out, $d_{v_i}(s, r)$ reaches its minimum value. $l(v_i, r)$ is constant in G, then $d(s, v_i)$ is the shortest path distance from s to v_i, that is $d(s, v_i) = d_{s, v_i}$.

Therefore, we have proved the shortest distance from s to t can be retrieved by the generalized A^* algorithm (denote as A^*-G algorithm) with arbitrary r. Meanwhile, if representative node r is selected as t, the algorithm is equivalent to the classic A^* algorithm; If the representative node r is selected as s, the algorithm is reduced to a special *Dijkstra's* algorithm. Moreover, A^*-G algorithm can return any node's shortest path during its search. And this is the basis of our *1-N A^** algorithm (denote as A^*-*1N* algorithm).

Now we present our A^*-*1N* algorithm. Given a non-negative graph $G(V, E)$, a starting node $s \in V$ and a set of target nodes $T \subset V$, A^*-*1N* algorithm aims to answer the batch shortest path query $Q_{s,T}$. According to the A^*-G algorithm, each query $q_{s,t} \in Q_{s,T}(t \in T)$ can be answered by running the algorithm once with its corresponding r. Since $r \in V$ can be arbitrarily selected, one fixed r can be used for all the queries $q_{s,t} \in Q_{s,T}(t \in T)$, as long as these queries have the same starting node s. The algorithm terminates when all nodes $v_i \in T$ have popped out and $Q_{s,T}$ has been answered completely. As for the *M-1* query, we can solve it reversely.

3.2 Representative Node Selection

It should be noted that although the representative node r can be arbitrarily chosen from V, its location affects the algorithm performance dramatically. Recall that $d_v(s, r)$ of the nodes popping out from H are in no descending order. Therefore, if node v_n is the last node popping out, it has the largest value of $d_{v_n}(s, r)$. Then any node v_i with $d_{v_i}(s, r) \leq d_{v_n}(s, r)$ must have popped out from H before the termination of algorithm. In another word, the location of r determines the way the search space grows. Only after every $t \in T$ are covered by this space can the search terminate. Thus, limiting the searched node number and search space is equivalent to lowering $d_{v_n}(s, r)$. So the representative node selection is transfered to an optimization problem as follows:

$$minimize(max\{d_{v_i}(s, r)\}), \quad subject\ to\ v_i \in T, r \in V$$

Suppose v_m is a node in T with largest shortest path length from s. Since minimizing the maximum value of $d_{v_i}(s, r)$ is our objective, and $d_{v_i}(s, r) = d_{s,v_i} + l(v_i, r)$ when v_i popped out of H. It is easy and direct to just minimize $l(v_m, r)$ to zero, and then r is exactly v_m. But selecting r as v_m cannot always minimize the maximum value of $d_{v_i}(s, r), v_i \in T$, as v_m may not be the last node popped out from H. It is possible that $d_{s,v_n} + l(v_n, v_m) > d_{s,v_m} + 0$ when selecting v_m as the representative node, where v_n is the last node popped out. Then v_m is not the optimal representative node. As v_m cannot be determined before the query start, v_m can be approximated to node $v_{farthest}$ with the largest euclidean distance from the starting node. In consequence, $v_{farthest}$ is the suboptimal representative node, which will be verified in Sect. 4.

The above selection is based on the premise that the target nodes are geographically close. However, the difference of $d_{s,v_n} + l(v_n, v_{farthest})$ and $d_{s,v_{farthest}}$ may be relatively big when target set is wildly distributed on G. Therefore, it would result in a larger searching space to cover all the target nodes and one single representative node won't be effective enough in controlling the searching space. Moreover, the *A*-1N* algorithm may even lose its advantages compared with the *Dijkstra's*. Hence, it is natural to decompose a large target set into several small clusters and run the *A*-1N* algorithm on each of them.

Algorithm 1. Angle-Based Decomposition

Input: $G(V, E)$, Starting node $s \in V$, Target node set $T \in V$
Output: $\{T_i\}, i = 0, \cdots, k$ with $\cup T_i = T$

1   begin
2   for $t \in T$ do
3       $H_1.insert(t, angle(s, t))$ // H_1 is a priority queue ordered by $angle(s, t)$ in ascending order
4     $cID \leftarrow 0$
5     $minAngle \leftarrow 0$
6     $creatNewCluster \leftarrow true$
7   while H_1 is not empty do
8       $v \leftarrow H_1.pop()$
9   if $creatNewCluster$ is false then
10   if $angle(s, v) \leq minAngel + \theta$ then
11           $T_{cID}.add(v)$
12   else
13           $cID \leftarrow cID + 1$
14           $creatNewCluster \leftarrow true$
15   if $creatNewCluster$ is true then
16         $T_{cID} \leftarrow \phi$
17         $T_{cID}.add(v)$
18         $minAngle \leftarrow angle(s, v)$
19         $creatNewCluster \leftarrow false$
20   return $\{T_i\}, i = 0, \cdots, k$

3.3   Decomposed A*-$1N$ Algorithm

In this section, we propose two decomposition approaches to solve the performance deterioration problem when T is sparse. First of all, we use $angle(s, v)$ to denote the included angle between vector (s, v) and $(1, 0)$. If it is more efficient to answer $q(s, v_1)$ and $q(s, v_2)$ separately rather than together when the angle difference of two nodes $angle(s, v_1)$ and $angle(s, v_2)$ is larger than θ, then θ is called the *decomposition angle threshold*. And it is reasonable to decompose a target node set T into several small clusters if the T is widespread.

After we get the cluster set $\{T_i\}$, we can either run the A*-$1N$ algorithm sequentially or in parallel. The number of cluster is bounded by $\lceil \frac{360}{\theta} \rceil$, which is much smaller compared to N. Therefore, it has a much smaller number of threads than running N times A* algorithm, which could save loads of system resources. Now we describe the angle-based and distance-based approaches in the following sections.

Angle-Based Decomposition Method. As the name indicates, the *angle-based* method only considers the included angles of the target nodes. It would create a set of clusters whose largest included angles are all smaller than the threshold. Firstly, we visit the target nodes in the angle-increasing order. If the included angle between the current node v_i and the first node v_0 is smaller than θ, we put v_i into the same cluster of v_0. Otherwise, if the included angle is larger than θ, we finish creating the cluster $T_0 = \{v_0, \cdots, v_{i-1}\}$ and use v_i as the first node in the second cluster. This procedure runs on until all the target nodes are processed. In the end, we get a set of target node clusters. The details of the method is shown in Algorithm 1. We use a min-heap H_1 to implement the sorting. The time complexity is $O(|T| \log |T|)$.

Distance-Based Decomposition Method. The distance-based decomposition method is based on the observation of the A*-$1N$ algorithm's searching space shape, which is roughly elliptical. Obviously, it is the farthest node that determines the size of this ellipse (on the far end of the long axis). Therefore, the total searching space could be saved if the farthest node from s locates in the middle of a decomposed cluster (along the long axis). Otherwise, we have to extend the searching space in order to cover it. The details is shown in Algorithm 2.

The target nodes are first sorted by their distances to s. Then we traverse the nodes from far to near. For the first node, we create a cluster T_0 to contain it. For the latter node v with largest distance among the left nodes, the minimum angle difference between v and existing clusters is calculated. If the minimum angle difference is no larger than $0.5 * \theta$, v is then added into one existing cluster with the smallest angle difference. Otherwise, a new cluster is created to contain v. This process won't stop until all target nodes are assigned to one of the clusters. We also implement it using max-heap, so the time complexity is $O(|T| \log |T|)$.

Algorithm 2. Distance-Based Decomposition

Input: $G(V, E)$, Starting node $s \in V$, Target node set $T \in V$
Output: $\{T_i\}, i = 0, \cdots, k$ with $\cup T_i = T$

```
1  begin
2  |   for t ∈ T do
3  |   |   H₂.insert(t, angle(s, t), l(s, t)) //H₂ is a priority queue ordered by l(s, t) in
   |   |   ascending order
4  |   while H₂ is not empty do
5  |   |   v ← H₂.pop()
6  |   |   //C is the existing cluster set
7  |   |   for cluster Tᵢ ∈ C do
8  |   |   |   minAngleDiff = minangle(v, Tᵢ)
9  |   |   if minAngleDiff is no larger than 0.5 * θ then
10 |   |   |   add v in existing cluter closest to v
11 |   |   else
12 |   |   |   construct a new cluster Tⱼ = ∅
13 |   |   |   Tⱼ.add(v)
14 |   |   |   C.add(Tⱼ)
15 |   return cluster set C = {Tᵢ}
```

3.4 Batch Query Answering

The batch shortest path query can be represented as the *M-N* shortest path query, where M is the number of source nodes and N is the number of target nodes. Depending on the actual queries, it is easy to divide them into several *1-N* queries that share the similar source, and several *M-1* queries that share the similar target. For each *1-N* query, we further decompose it into n clusters, where $n \ll N$. As for the *M-1* query, it is the reverse version of *1-N*. In this way, the massive batch shortest path query can be answered by a set of decomposed *A*-1N* algorithms. It has a much smaller thread number than the pairwise *A**, and is much faster than the *Dijkstra*'s.

4 Experiments

We test our methods on *Beijing* road network which consists of 302,364 nodes and 387,588 roads. Although the map is 185 km × 178 km, only the central area of 120 km × 108 km is populated with many nodes. We use this zone in the experiments. We randomly select a node in the central area as the origin s of 1-N shortest path queries, and generate target set randomly from different directions around s. Five target sets are chosen in each direction and we report the average performance.

Our algorithms are evaluated under different scenarios by changing the following factors: (1) the location of the representative node; (2) target set size n, namely number of nodes contained in the target set; (3) target box size e^2, namely size of the minimum bounding rectangle of the target set; (4) distance d from s to target set; and (5) target set distribution. The default values of n, d and e^2 are set to 20, 15 km, and 2 km × 2 km respectively in the experiments. We

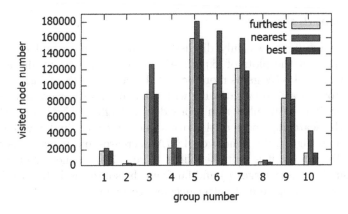

Fig. 1. Representative node selection

compare our algorithms with two alternatives, i.e., Dijkstra and A^*-N (running A^* algorithms for n times), in terms of *the number of visited nodes* denoted as $|\hat{V}|$. $|\hat{V}|$ represents the size of search space which determines running time. The higher the $|\hat{V}|$, the worse the performance. All experiments are carried out on an Intel Core i7-7500U CPU 2.9 GHz with 8 GB RAM.

4.1 Location of the Representative Node r

Figure 1 shows the results of 10 groups of experiments. Each group randomly selects a target box and chooses 20 nodes in it. We compare the performance of selecting the *nearest node*, the *farthest node* and the *best-performed node* as the representative node r. Obviously, the *nearest node* always has the largest $|\hat{V}|$, while the *farthest* and the *best-performed* are much better and the difference between them is relatively small. Because it always takes N times to get the *best-performed* node and the *farthest* node is suboptimal, we use the *farthest* node as the representative node r in the following experiments.

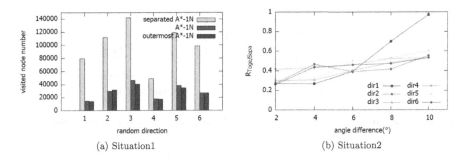

(a) Situation1   (b) Situation2

Fig. 2. Target set distribution

4.2   Effect of d, n and e

Due to the page limit, we only discuss the effects of d, n and e on A^*-$1N$, but do not present the plots of experiment result here. First of all, as the distance d increases, A^*-$1N$ gains more benefit. Because the searching space of the *Dijkstra*'s grows nearly quadratically, while that of A^*-$1N$ grows much slower. Secondly, target node number n in a box does not affect A^*-$1N$ and *Dijkstra*'s much, but the searching space of A^* grows linearly to n. Finally, when the box size e^2 becomes bigger, the benefit of A^*-$1N$ decreases and it performs closer to *Dijkstra*'s. To sum up, it is more beneficial to run A^*-$1N$ when the target set is compact (smaller e and larger n) and far from the source node (larger d). When the targets are sparse, we had better decompose them.

4.3   Target Set Distribution

The target set distribution covers a wide range of parameters, such as the dispersion degree of target set relative to start node, distortion of target set, the shape of target set and so on. We consider the distortion of target set tentatively.

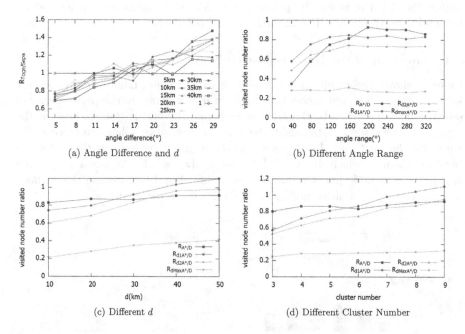

(a) Angle Difference and d

(b) Different Angle Range

(c) Different d

(d) Different Cluster Number

Fig. 3. Performance comparison of *1-N A^**, *Decomposed A^** and *Dijkstra*'s algorithm

A distortion target set is made up of several small boxes with close distance. We test two totally different and extreme situations. In the first one, boxes are along the same direction but different distance with respect to start node. In the second one, the direction along which the boxes locate, is almost vertical with the direction from start node to the boxes. We set 4 small target boxes in each target set with equal angle difference α between every two adjacent boxes. 5 groups of sets, with different $\alpha(2°, 4°, 6°, 8°, 10°)$, are randomly selected from 6 direction respectively. These 6 directions distribute uniformly around the start node. And separated A^*-$1N$ algorithm is compared with A^* algorithm and the *Dijkstra*'s algorithm. Figure 2(a) shows the result of situation 1. The outermost A^*-$1N$ uses the farthest box as input. Obviously, the visited node number of A^*-$1N$ and outermost A^*-$1N$ are fairly close. Nodes in other three closer boxes will also be searched in outermost A^*-$1N$ algorithm as long as their box sizes are not too large. So when the target nodes locate in the range of certain direction, it is the farther box that determines algorithm efficiency rather than those closer ones. Figure 2(b) shows the experiment result of situation two. $R_{Togo/Sepa}$ denotes the ratio of $|\hat{V}|$ in A^*-$1N$ and in separated A^*-$1N$. It can be seen that $R_{Togo/Sepa}$ increases in each direction besides some fluctuations. As the angle difference between the adjacent boxes increases, target set distributes wider. The search area of A^*-$1N$ algorithm becomes broader to cover all the target nodes, while the separated A^*-$1N$ algorithm does not change too much. So A^*-$1N$ loses its advantage against the wide-separated target set.

4.4   Decomposed Batch Shortest Path

In order to determine the value of the decompose angle threshold θ, we randomly select two nodes with different angle differences with the same distance d from start node. We test 8 groups with d ranging from 5 km to 40 km. The result is shown in Fig. 3(a). When $R_{Togo/Sepa}$ reaches 1, the corresponding angle difference is obtained as θ. Although there are fluctuations in Fig. 3(a), the value of θ trends larger as d increases.

The two decomposition methods are compared with *Dijkstra*'s algorithm and A^*-$1N$ algorithm. The ratio of $|\hat{V}|$ in A^*-$1N$, *angle-based decomposition* A^*-$1N$ and *distance-based decomposition* A^*-$1N$ algorithm to *Dijkstra*'s algorithm are denoted as $R_{A^*/D}$, $R_{d1A^*/D}$ and $R_{d2A^*/D}$, respectively. And the ratio of the maximum $|\hat{V}|$ among decomposed clusters to that of Dijkstra's algorithm is denoted as $R_{dMaxA^*/D}$. We test the angle range of target set distribution, the maximum distance d between start node and target set, and the generated cluster number. The default value of angle range, d and cluster number are π, 3 km and 5, respectively.

In the first experiment, angle range differs from $40°$ to $320°$. In the second experiment, the value of d ranges from 10 km to 50 km. In the third experiment the box number variates from 3 to 9. In each experiment, 100 groups of target cluster is generated under each parameter combination to fully reflect the performance of each algorithm. And the experiment results are shown in Figs. 3(b), (c) and (d) respectively.

In Fig. 3(b), $R_{A*/D}$, $R_{d1A*/D}$ and $R_{d2A*/D}$ all tend to increase, especially for $R_{A*/D}$. Target nodes distribute more dispersedly with larger angle range, which results in a wider search space of $A*$-$1N$ algorithm. Target nodes set will be decomposed into more clusters, and $|\hat{V}|$ of two decomposition method will certainly increase, which causes the increase of $R_{decomp1/D}$ and $R_{decomp2/D}$. And they are both smaller than $R_{A*/D}$ when angle range is more than $200°$, which means that it is better to answer batch query by decomposed method than 1-N $A*$ algorithm when target nodes are more widespread. Figure 3(c) shows that $R_{A*/D}$ is relatively steady, while $R_{d1A*/D}$ and $R_{d2A*/D}$ increase with larger d. Longer distance from the start node and target nodes results in larger search space for every decomposed target set, which is the reason for the growth of $R_{d1A*/D}$ and $R_{d2A*/D}$. In Fig. 3(d), $R_{d1A*/D}$ and $R_{d2A*/D}$ grow along with the increase of cluster number, and $R_{A*/D}$ rises slightly. The distribution of target nodes is more uniform when the generated cluster number is larger. And the decomposition will result in smaller clusters, which increases the cost of computation after decomposition. So it will be more efficient to decompose the target set when nodes in it distribute closely locally. It is common in Fig. 3 that $R_{decomp2/D}$ is always smaller than $R_{decomp1/D}$, which means distance-based decomposition method is more efficient.

It should be noted that the result of $R_{A*/D}$, $R_{d1A*/D}$ and $R_{d2A*/D}$ are all based on sequential computation. If parallel computation is applied, only the decomposed cluster which consumes the maximum cost needs to be taken into consideration. And the value of $R_{dMaxA*/D}$ is smaller than other three ratios and stay stable under different situations. So if parallel computation is permitted, batch query efficiency will be improved greatly by decomposition method.

5   Conclusion

In this paper, we have proposed a $A*$-$1N$ algorithm and two decomposition methods to solve the massive shortest path query problem by reusing sharable computation. We first generalize the classic $A*$ algorithm to answer shortest path queries with N targets. After that, we studied the effectiveness of different representative node location. In order to cope with the situation where the target node set is sparse, we come up with two effective approaches to decompose a large target set into several smaller clusters. Our extensive experiments have fully tested the performance factors and confirmed that our method is effective in saving the computation resources when confronting a swarm of queries. This work is the basis of the future *batch M-N shortest path* problem.

Acknowledgment. This research is partially supported by the Australian Research Council (Grants No. DP150103008 and DP170101172).

References

1. Goldberg, A.V., Harrelson, C.: Computing the shortest path: a search meets graph theory. In: Proceedings of the Sixteenth Annual ACM-SIAM Symposium on Discrete Algorithms, pp. 156–165. Society for Industrial and Applied Mathematics (2005)
2. Geisberger, R., Sanders, P., Schultes, D., Delling, D.: Contraction hierarchies: faster and simpler hierarchical routing in road networks. In: McGeoch, C.C. (ed.) WEA 2008. LNCS, vol. 5038, pp. 319–333. Springer, Heidelberg (2008). https://doi.org/10.1007/978-3-540-68552-4_24
3. Möhring, R.H., Schilling, H., Schütz, B., Wagner, D., Willhalm, T.: Partitioning graphs to speedup Dijkstra's algorithm. J. Exp. Algorithm. (JEA) **11**, 2–8 (2007)
4. Cohen, E., Halperin, E., Kaplan, H., Zwick, U.: Reachability and distance queries via 2-hop labels. SIAM J. Comput. **32**(5), 1338–1355 (2003)
5. Akiba, T., Iwata, Y., Yoshida, Y.: Fast exact shortest-path distance queries on large networks by pruned landmark labeling. In: Proceedings of the 2013 ACM SIGMOD International Conference on Management of Data, pp. 349–360. ACM (2013)
6. Ouyang, D., Qin, L., Chang, L., Lin, X., Zhang, Y., Zhu, Q.: When hierarchy meets 2-hop-labeling: efficient shortest distance queries on road networks. In: Proceedings of the 2018 International Conference on Management of Data, pp. 709–724. ACM (2018)
7. Samet, H., Sankaranarayanan, J., Alborzi, H.: Scalable network distance browsing in spatial databases. In: Proceedings of the 2008 ACM SIGMOD International Conference on Management of Data, pp. 43–54. ACM (2008)
8. Wang, S., Xiao, X., Yang, Y., Lin, W.: Effective indexing for approximate constrained shortest path queries on large road networks. Proc. VLDB Endow. **10**(2), 61–72 (2016)
9. Wang, S., Lin, W., Yang, Y., Xiao, X., Zhou, S.: Efficient route planning on public transportation networks: a labelling approach. In: Proceedings of the 2015 ACM SIGMOD International Conference on Management of Data, pp. 967–982. ACM (2015)
10. Dijkstra, E.W.: A note on two problems in connexion with graphs. Numer. Math. **1**(1), 269–271 (1959)
11. Cooke, K.L., Halsey, E.: The shortest route through a network with time-dependent internodal transit times. J. Math. Anal. Appl. **14**(3), 493–498 (1966)
12. Hart, P.E., Nilsson, N.J., Raphael, B.: A formal basis for the heuristic determination of minimum cost paths. IEEE Trans. Syst. Sci. Cybern. **4**(2), 100–107 (1968)
13. Li, L., Hua, W., Du, X., Zhou, X.: Minimal on-road time route scheduling on time-dependent graphs. Proc. VLDB Endow. **10**(11), 1274–1285 (2017)
14. Li, L., Zheng, K., Wang, S., Hua, W., Zhou, X.: Go slow to go fast: minimal on-road time route scheduling with parking facilities using historical trajectory. VLDB J. Int. J. Very Large Data Bases **27**(3), 321–345 (2018)
15. Goldberg, A.V., Kaplan, H., Werneck, R.F.: Reach for a: efficient point-to-point shortest path algorithms. In: Proceedings of the Meeting on Algorithm Engineering & Experiments, pp. 129–143. Society for Industrial and Applied Mathematics (2006)
16. Wagner, D., Willhalm, T., Zaroliagis, C.: Geometric containers for efficient shortest-path computation. J. Exp. Algorithm. (JEA) **10**, 1–3 (2005)

17. Fu, A.W.-C., Wu, H., Cheng, J., Wong, R.C.-W.: IS-LABEL: an independent-set based labeling scheme for point-to-point distance querying. Proc. VLDB Endow. **6**(6), 457–468 (2013)
18. Li, Y., Yiu, M.L., Kou, N.M., et al.: An experimental study on hub labeling based shortest path algorithms. Proc. VLDB Endow. **11**(4), 445–457 (2017)
19. Jiang, M., Fu, A.W.-C., Wong, R.C.-W., Xu, Y.: Hop doubling label indexing for point-to-point distance querying on scale-free networks. Proc. VLDB Endow. **7**(12), 1203–1214 (2014)
20. Sankaranarayanan, J., Alborzi, H., Samet, H.: Efficient query processing on spatial networks. In: Proceedings of the 13th Annual ACM International Workshop on Geographic Information Systems, pp. 200–209. ACM (2005)
21. Qi, Z., Xiao, Y., Shao, B., Wang, H.: Toward a distance oracle for billion-node graphs. Proc. VLDB Endow. **7**(1), 61–72 (2013)
22. Reza, R.M., Ali, M.E., Hashem, T.: Group processing of simultaneous shortest path queries in road networks. In: 2015 16th IEEE International Conference on Mobile Data Management (MDM), vol. 1, pp. 128–133. IEEE (2015)
23. Mahmud, H., Amin, A.M., Ali, M.E., Hashem, T., Nutanong, S.: A group based approach for path queries in road networks. In: Nascimento, M.A., et al. (eds.) SSTD 2013. LNCS, vol. 8098, pp. 367–385. Springer, Heidelberg (2013). https://doi.org/10.1007/978-3-642-40235-7_21

Extracting Temporal Patterns from Large-Scale Text Corpus

Yu Liu[✉], Wen Hua, and Xiaofang Zhou

School of Information Technology and Electrical Engineering,
The University of Queensland, Brisbane, Australia
{yu.liu,w.hua}@uq.edu.au, zxf@itee.uq.edu.au

Abstract. Knowledge, in practice, is time-variant and many relations are only valid for a certain period of time. This phenomenon highlights the importance of designing temporal patterns, i.e., indicating phrases and their temporal meanings, for temporal knowledge harvesting. However, pattern design is extremely laborious and time consuming even for a single relation. Therefore, in this work, we study the problem of temporal pattern extraction by automatically analysing a large-scale text corpus with a small number of seed temporal facts. The problem is challenging considering the ambiguous nature of natural language and the huge amount of documents we need to analyse in order to obtain highly representative temporal patterns. To this end, we introduce various techniques, including corpus annotation, pattern generation, scoring and clustering, to reduce ambiguity in the text corpus and improve both accuracy and coverage of the extracted patterns. We conduct extensive experiments on real world datasets and the experimental results verify the effectiveness of our proposals.

Keywords: Temporal knowledge harvesting · Temporal patterns · Text mining

1 Introduction

With the technological advancements in Information Extraction (IE), large-scale Knowledge Bases (KBs) have been constructed for semantic understanding, Question Answering (QA) and other advanced tasks. KBs such as DBpedia [2], TextRunner [28], NELL [18], Probase [27], and YAGO [16], are built automatically from unstructured text by extracting millions of entities and relational facts. However, most of these KBs regard relation instances as time-invariant and ignore the corresponding valid temporal period. Actually, many relations are changing and involving over time, i.e. they are only valid for a certain time period. For example, the relation instance *SpouseOf*("Brad Pitt", "Angelina Jolie") holds true only over the temporal period of 2014 to 2016. The temporal scope of relations is particularly important and beneficial in many application scenarios including QA systems, text summarisation, timeline generation, etc. [5].

© Springer Nature Switzerland AG 2019
L. Chang et al. (Eds.): ADC 2019, LNCS 11393, pp. 17–30, 2019.
https://doi.org/10.1007/978-3-030-12079-5_2

Research on complementing KBs with a temporal dimension is very current. To the best of our knowledge, only Freebase [3] and Yago2 [13], two recently constructed large KBs, have timestamped facts. However, Freebase is a collaborative KB constructed mainly by its community. Yago2 extracts temporal facts using regular expressions only from Wikipedia Infoboxes, which limits its coverage and applicability to widely available free texts [22]. Although extracting temporal facts from free texts has been studied, including T-Yago [26], PRAVDA [25], CoTS [22], TIE [15], it is still limited to specific domains and cannot be applied to large-scale KBs.

Inspired by the pattern-based approach for constructing large-scale generalised KBs [27], we resort to temporal patterns for temporal knowledge extraction. For example, if we know the phrase "get married" strongly implies the beginning of a marriage, then we can apply the temporal pattern ($PERSON$, $PERSON$, "get married", $TIME$) on the text corpus to extract the start date of all possible facts of the relation $SpouseOf$. However, it is extremely laborious and time-consuming to manually construct temporal patterns even for a single relation. Hence, in this work, we focus on designing automatic methods for extracting temporal patterns, in particular indicating phrases (e.g., "get married") and their temporal status (e.g., $START$ and END), from large-scale text corpus.

Given a seed set of facts between entities (e.g., "Brad Pitt" and "Angelina Jolie") along with their valid time (e.g., [2014, 2016]), our algorithm searches in the text corpus for sentences that contain both entities and time expressions (e.g., "Pitt has been married to actress Angelina Jolie since 2014", "In September 2016, Jolie filed for divorce from Pitt", etc.) and then extracts and aggregates temporal patterns (e.g., ("be married", $START$), ("file for divorce", END), etc.) from these sentences. Although the idea sounds simple, challenges still abound. First, the text corpus is noisy and informal, full of nicknames, abbreviations, spelling mistakes, pronouns, ambiguous entities, etc., which greatly limits the number of sentences that can be retrieved for each seed. Second, different phrases can indicate the existence of a relation to different extent. Given a set of retrieved sentences, it is non-trivial to select indicating phrases and determine their indicating strength. Finally, the text corpus is extremely large, making it infeasible to traverse the entire corpus when extracting temporal patterns for a relation. To address these issues, our major contributions in this work can be summarised as below:

- We demonstrate the importance of temporal knowledge harvesting, and propose a novel framework for automatic temporal pattern extraction from a text corpus.
- We introduce various techniques, including corpus annotation, pattern generation, scoring and clustering, to reduce ambiguity in the text corpus and improve both the accuracy and coverage of the extracted patterns.
- We conduct extensive experiments on real world datasets and the experimental results verify the effectiveness of our proposed framework.

The rest of this paper is organised as follows: in Sect. 2, we formally define the problem of temporal pattern extraction and introduce proposed techniques in detail; our experimental results are presented in Sect. 3, followed by a summary of related work on temporal knowledge extraction in Sect. 4 and a brief conclusion in Sect. 5.

2   Temporal Pattern Extraction

Definition 1 (Temporal Fact). *A temporal fact, denoted as $r(e_1, e_2, [t_s, t_e])$, indicates the existence of a relation r between entities e_1 and e_2 during the time period $[t_s, t_e]$. For example, the temporal fact SpouseOf("Brad Pitt", "Angelina Jolie", [2014, 2016]) means there is a marriage relation between entities "Brad Pitt" and "Angelina Jolie", and it starts in 2014 and terminates in 2016.*

Definition 2 (Temporal Pattern). *We define the temporal pattern p as a phrase that can, to some extent, imply the commencement or termination of a relation r. Specifically, p consists of two parts: an **indicating phrase** v (e.g., verb phrase) and its **temporal status** $sta \in \{START, END\}$, i.e., $p = (v, sta)$. For example, given relation SpouseOf, its temporal pattern could be ("get married", START), ("get divorced", END), and ("hold a wedding", START), etc. Obviously, different phrases can indicate the temporal status of relation r to different extent. Therefore, we use $w(p)$ to represent the indicating strength of pattern p.*

Definition 3 (Temporal Pattern Extraction). *Given a text corpus $\mathbb{D}$ and a seed set of temporal facts $\{r(e_1, e_2, [t_s, t_e])\}$, we aim at extracting a collection of weighted temporal patterns, i.e., $\{< p, w(p) >\}$ that can indicate relation r.*

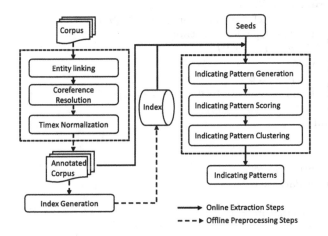

Fig. 1. Framework overview.

Figure 1 illustrates our proposed framework for temporal pattern extraction, which consists of two parts: offline corpus preprocessing and online pattern extraction. Offline preprocessing is important for improving accuracy and efficiency of our algorithm. As introduced in Sect. 1, we need to collect all sentences containing the seed facts (i.e., entities and time expressions) for temporal pattern extraction. However, the text corpus is noisy and informal, full of nicknames, abbreviations, pronouns, ambiguous entities, etc., which greatly limits the number of sentences that can be retrieved for each seed and lowers down the accuracy and coverage of extracted patterns. Therefore, we annotate the underlying text corpus using entity linking (TAGME [11]), co-reference resolution (Stanford Neural Co-reference [7]) and time expression normalisation (HeidelTime [20]) during the offline stage. Furthermore, considering the large scale of the text corpus, we construct an inverted-list-like index to avoid scanning the entire text corpus and hence speed up online pattern extraction. During the online stage, we handle input seeds one-by-one to generate temporal patterns (Sect. 2.1). We then propose various weighting strategies to estimate the indicating strength of each pattern, and aggregate patterns among all the input seeds

---

Algorithm 1. Online Temporal Pattern Extraction.

Input: a seed set $\{r(e_1, e_2, [t_s, t_e])\}$, an annotated text corpus $\mathbb{D}$, an inverted index $I_{e,t}$

Output: weighted temporal patterns $\mathbb{P}_r$ for relation r

1   $\mathbb{P}_r = \emptyset$
2   **foreach** $r(e_1, e_2, [t_s, t_e])$ **do**
3      $\mathbb{S}$=RetrieveRelevantSentences($e_1, e_2, [t_s, t_e], \mathbb{D}, I_{e,t}$)
4   **foreach** $s \in \mathbb{S}$ **do**
5         $\mathbb{V}_s$=ExtractVerbPhrases(s)
6         $\mathbb{T}_s$=ExtractTimeExpressions(s)
7         $tree_s$=ExtractParseTree(s)
8   **foreach** $v \in \mathbb{V}_s$ **do**
9   t_v=FindRelatedTimeExpression($v, \mathbb{T}_s, tree_s$)
10   **if** $(|t_v - t_s| < |t_v - t_e|$ **then**
11              $p = (v, START)$
12   **else**
13              $p = (v, END)$
14           $w_s(p)$=ScorePattern($p, s, e_1, e_2, [t_s, t_e]$)
15           $\mathbb{P}_r = \mathbb{P}_r \cup \{\langle p, w_s(p) \rangle\}$
16   **end**
17   **end**
18 **end**
19 AggregatePatternScore($\mathbb{P}_r$)
20 $\{\mathbb{P}_c\}$=ClusterPattern($\mathbb{P}_r$)
21 **foreach** $\mathbb{P}_c$ **do**
22   AdjustPatternScore($\mathbb{P}_c$)
23 **end**
24 Return $\{\mathbb{P}_r\}$

(Sects. 2.2 and 2.3). Due to space limit, we only introduce some major techniques of online pattern extraction in this paper, as illustrated in Algorithm 1.

2.1 Temporal Pattern Generation

Recall that temporal patterns are phrases that can indicate the commencement or termination of a relation to some extent. We regard verb phrases in a relevant sentence as the candidate temporal patterns since the sentence is talking about an event that can imply the existence of the target relation. In this work, we consider a sentence s as relevant to a seed $r(e_1, e_2, [t_s, t_e])$ if and only if s satisfies the following conditions:

- s contains both e_1 and e_2;
- s contains at least one time expression t which is temporally close to either t_s or t_e.

For each relevant sentence s, we extract verb phrases v and time expressions t based on the definition of verb phrase in Open IE systems [10] and the definition of time expression in the Timex3 annotation of TimeML [20], respectively. The temporal status of a phrase $sta \in \{START, END\}$ can be determined by its corresponding time expression. That is, a phrase is highly possible to indicate the commencement (resp. termination) of a relation if its time expression is close to the start date t_s (resp. end date t_e) of the input seed (lines 9–13 in Algorithm 1). However, multiple verb phrases and time expressions might coexist in a sentence, and matching verb phrases to relevant time expressions is not easy. A straightforward solution is to consider the distance (i.e., number of words) between v and t in sentence s, but it fails sometimes. Consider the sentence "Pitt met Friends actress Jennifer Aniston in 1998 and married her in a private wedding ceremony in Malibu on July 29, 2000." as an example. The most relevant time expression for verb "marry" obtained using the naive distance-based method is "1998" rather than "July 29, 2000". In fact, humans can correctly identify the relatedness between phrases and time expressions in a sentence since they understand the syntactic structure of that sentence. Therefore, we resort to **parse tree** to locate relevant time expressions.

Figure 2 illustrates an example parse tree. In the parse tree, leaf nodes are tokens in the sentence and internal nodes are their labels, i.e. part-of-speech tags. Intuitively, a verb phrase v is more related to a time expression t if v is closer to t in the parse tree. Hence, given a pair of (v, t), we define their tree distance $dist_s(v, t)$ as the length (i.e., number of edges) of path to traverse from v and t to their lowest common ancestor (LCA) in the parse tree. For example, $dist_s(\text{"marry"}, \text{"1998"}) = 3$ while $dist_s(\text{"marry"}, \text{"July 29, 2000"}) = 2$. We then regard the time expression that minimises $dist_s(v, t)$ as the most related time expression t_v of phrase v.

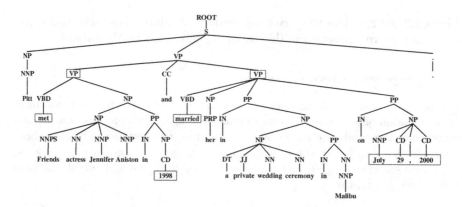

Fig. 2. Parse tree of sentence "Pitt met Friends actress Jennifer Aniston in 1998 and married her in a private wedding ceremony in Malibu on July 29, 2000."

2.2 Temporal Pattern Scoring

Obviously, temporal patterns have different indicating strength. For example, ("marry", $START$) has a much stronger indication than {("meet", $START$) for the beginning of relation $SpouseOf$. Moreover, the text corpus is intrinsically informal and noisy, which means we cannot guarantee that all the temporal patterns extracted from the text corpus indicate a relation. However, as long as these patterns can be scored and ranked effectively, those top-k patterns are still invaluable for temporal knowledge harvesting. Therefore, it is essential to design an effective weighing scheme which can correctly estimate the indicating ability of each temporal pattern. In this work, we propose several heuristics to achieve a reasonable pattern scoring. Specifically, we regard a pattern $p = (v, sta)$ as indicative of relation r if and only if

- for a seed $r(e_1, e_2, [t_s, t_e])$ and a relevant sentence s that generates p, (1) v is close to either e_1 or e_2 in s; (2) v is close to its related time expression t_v in the parse tree of s; (3) t_v is temporally close to either t_s or t_e;
- p can be extracted from many sentences relevant to seed $r(e_1, e_2, [t_s, t_e])$;
- p can be generated by many seed facts of relation r.

In-Sentence Scoring. We first score pattern p at sentence level based on the first heuristic. In particular, given a seed fact $r(e_1, e_2, [t_s, t_e])$ and a relevant sentence s, we calculate the weight of p in s, denoted as $w_s(p)$, using Eq. 1.

$$w_s(p) = \alpha \cdot w_s(v, e_1, e_2) + \beta \cdot w_s(v, t_v) + \gamma \cdot w_s(t_v, t_s, t_e) \tag{1}$$

In Eq. 1, $w_s(v, e_1, e_2)$, $w_s(v, t_v)$, and $w_s(t_v, t_s, t_e)$ capture the position of v in s, the correlation between v and t_v, and the temporal closeness between t_v and $[t_s, t_e]$, respectively. α, β, and γ are three parameters to reflect the relative importance of these features in the scoring function such that $\alpha + \beta + \gamma = 1$.

In particular, measuring temporal distance between two time expressions $|t_1 - t_2|$ is not straightforward. We observe that a time expression in practice is usually represented as either a specific date (e.g. "2006-01-0") or a temporal range (e.g. "2006") which, however, cannot be compared directly. Table 1 summaries all possible combinations of two time expressions t_1 and t_2, where star and horizontal bar represent a date and a temporal range respectively. Note that the last case where t_1 and t_2 have a partial overlapping does not occur due to time expression normalisation in the text corpus. To address this issue, we propose a unified method to calculate $|t_1 - t_2|$. In particular, each time expression t is transformed into a temporal range $[t.min, t.max]$ such that $t.min$ (resp. $t.max$) is the minimum (resp. maximum) possible date of t. $t.min = t.max$ if t is originally represented as a date. For instance, "May 2006" will be normalised as ["2006-05-01", "2006-05-31"]. Then we calculate $|t_1 - t_2|$ using the following equation.

$$|t_1 - t_2| = \begin{cases} 0, (t_1.min - t_2.min) \cdot (t_1.max - t_2.max) \leq 0 \\ \frac{|t_1.min - t_2.min| + |t_1.max - t_2.max|}{2}, otherwise \end{cases} \quad (2)$$

In Eq. 2, $(t_1.min - t_2.min) \cdot (t_1.max - t_2.max) \leq 0$ reflects that one time expression is fully contained in another, and their temporal distance is regarded as 0 in this work. Table 1 shows some examples of temporal distances $|t_1 - t_2|$.

Cross-Sentence Scoring. After pattern scoring at sentence level, we further aggregate all patterns to refine pattern weights. Aggregation heuristics follow the observation that a pattern becomes more indicative if it can be extracted from many relevant sentences and many seeds. Therefore, we aggregate pattern weights among all the sentences and seeds (TF-like method). Common phrases (e.g., "start to") are further eliminated (IDF-like method) and the scores are then normalised among all the patterns. We ignore the details in this paper due to space limit.

Table 1. Different cases and examples of temporal distances between two time expressions.

Case		Example	Distance (days)
t_1*	t_2*	"2006-03-01" vs. "2006-03-16"	15
t_1____	t_2*	"March 2006" vs. "2006-05-31"	75
t_1	t_2*	"2006" vs. "2006-03-01"	0
t_1__	__t_2	"March 2006" vs. "May 2006"	60
t_1 ___t_2		"2006" vs. "March 2006"	0
t_1 ___t_2		-	-

2.3 Temporal Pattern Clustering

We also observe that some of the temporal patterns are highly correlated in semantics, and hence their scores should be adjusted accordingly. For instance, the following two patterns ("get married", $START$) and ("hold a wedding", $START$) are semantically related. As a consequence, they are able to reciprocally promote each other whenever any of them is identified as highly indicative of a relation $SpouseOf$. We adopt word embedding techniques (e.g., Word2Vec [17]) to represent each phrase and apply existing density-based clustering method (e.g., DBSCAN [9]) to locate semantically related patterns, using the following distance function to find semantically related patterns.

$$dist(p, p') = 1 - cosine(vec(p), vec(p')) \tag{3}$$

Here, $vec(p)$ is the vector presentation of p obtained via word embedding on p's verb.

For each pattern cluster $\mathbb{P}_c$, we propose a **weighted-voting** method to adjust pattern scores in that cluster. In particular, we denote p^* as the highest ranked pattern in $\mathbb{P}_c$, namely, $p^* = \arg \max_{p_i \in \mathbb{P}_c} w(p_i)$. Then the final score of pattern p is adjusted as a combination of self-vote (whether p is originally indicative) and context-vote (whether p^* can semantically support p).

$$w(p) = w(p) + \delta \cdot sim(p, p^*) \cdot (w(p^*) - w(p)) \tag{4}$$

In Eq. 4, δ is an empirical decay factor (we set it as 0.8) to avoid negative effects, and $sim(p, p^*)$ is the semantic similarity between patterns p and p^*. We can see that the larger the similarity, the more support p can achieve from p^*.

3 Experiments

3.1 Experimental Settings

Datasets. In general, our framework can extract temporal patterns from any type of text corpus, such as web pages, news articles, tweets, etc., written in English. In this paper, we report our experimental results on the publicly available PRAVDA datasets [24, 25][1], which contain 23,000 soccer players' Wikipedia articles and around 110,000 online news articles mentioning players in "FIFA 100 list", as well as 88,000 news articles about persons mentioned in the "Forbes 100 list" and their Wikipedia articles.

Evaluation Metrics. To the best of our knowledge, there is no public gold-standard we can directly use to evaluate our algorithms for temporal pattern extraction. Therefore, we selected two relations (i.e., $SpouseOf$, $PlayForClub$) that are quite popular in existing temporal KBs [24, 25] and randomly sampled 20 seed facts for each relation from Wikipedia's InfoBox. We ran our pattern

---

[1] The datasets can be downloaded from https://www.mpi-inf.mpg.de/departments/ databases-and-information-systems/research/yago-naga/pravda/.

extraction algorithms on all the 40 (i.e., 2*20) seeds, and invited five annotators with different backgrounds to label the quality of the extracted patterns (from 1 to 5). The final label was based on the average of all the voting. In this way, we obtained a high quality pattern library $\mathbb{P}^*$. We then evaluated the extracted patterns under different settings using a widely-adopted ranking criterion $nDCG$ at the top-k patterns, namely $nDCG_k$.

Parameter Setting. The most important parameters employed in this work are α, β, and γ, which represent the corresponding contribution of $w_s(v, e_1, e_2)$, $w_s(v, t_v)$, and $w_s(t_v, t_s, t_e)$ to the overall pattern score (details introduced in Sect. 2.2). As we do not have enough benchmark to automatically learn these parameters, in the experiments, we iterated through all possible combinations of the parameters (i.e., range from 0 to 1 with a step of 0.1) and reported the best performance achieved.

3.2   Accuracy of Temporal Patterns

We evaluated the accuracy of our framework for temporal pattern extraction from the following five aspects: (1) the influence of relation type, (2) the influence of seed popularity, (3) the influence of seed set size, (4) the effectiveness of pattern scoring features, and (5) the effectiveness of pattern clustering for score adjustment.

Table 2. Pattern accuracy of different relations.

Relation	$START$ patterns				END patterns			
	$nDCG_1$	$nDCG_3$	$nDCG_5$	$nDCG_{10}$	$nDCG_1$	$nDCG_3$	$nDCG_5$	$nDCG_{10}$
SpouseOf	1.00	1.00	1.00	0.91	1.00	1.00	1.00	0.76
PlayForClub	1.00	1.00	0.82	0.88	0.60	0.54	0.50	0.48

The Influence of Relation Types. We evaluated the accuracy of our algorithms for different relation types. In particular, we extracted $START$ and END patterns based on all the 20 seeds for each relation, and reported $nDCG_k$ in Table 2. We can see that $START$ patterns are generally more accurate than END patterns which, we believe, is natural since humans usually talk more about the beginning rather than the termination of a relation. Furthermore, relation $SpouseOf$ performs consistently well for both $START$ and END patterns, while we can only obtain accurate $START$ patterns for relation $PlayForClub$. However, the overall performance of our algorithm is still satisfactory, especially for $START$ patterns. It is worth noting that $nDCG_k$ decreases gradually when k rises, which means most of the correct patterns are ranked relatively high (e.g., in top-1, top-3 or top-5). This also verifies the effectiveness of our proposed strategies for pattern generation and scoring. For the remaining of the experiments, we only report average accuracy of temporal patterns among both relations and status ($START$ and END).

The Influence of Seed Popularity. One important problem is how to select input seeds when applying our framework for temporal pattern extraction. In this work, we estimated seed quality by its popularity in the text corpus, and evaluated its influence on pattern extraction. In particular, we counted the number of sentences containing both e_1 and e_2 for each seed and regarded it as seed popularity. We then conducted pattern extraction given each single seed, and reported pattern accuracy in Fig. 3(a). The x-axis denotes seed popularity, which is divided into four intervals: $[0-10]$, $[11-50]$, $[51-500]$, and $[500+]$, and the y-axis is the $nDCG_k$ value. From Fig. 3(a) we can observe an overall improvement of accuracy when seed popularity increases. This is consistent with our expectation that popular seeds are more powerful in extracting temporal patterns than unpopular ones. Based on such an observation, we can select input seeds using popular entities, which is a natural way adopted in practice.

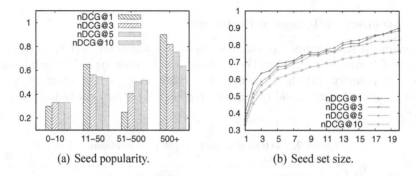

(a) Seed popularity.   (b) Seed set size.

Fig. 3. Pattern accuracy of different seed popularities and seed set sizes.

The Influence of Seed Set Size. We also evaluated the influence of seed set size on the performance of pattern extraction. Given a seed set size denoted as $|s|$, for each relation, we enumerated all possible combinations of $|s|$ seeds (i.e., $\binom{20}{|s|}$ combinations) and calculated the average $nDCG_k$ after applying each seed set for pattern extraction. We then averaged among both relations to obtain the $nDCG_k$ for the given seed set size $|s|$. Figure 3(b) reports the variation of $nDCG_k$ when $|s|$ ranges from 1 to 20. The results illustrate that pattern accuracy improves when $|s|$ increases. It is worth noting that all the 20 seeds for each relation were randomly sampled, and hence most seeds are not very popular (or of high quality). Therefore, it is not necessary to manually select popular seeds for pattern extraction. Even randomly sampled seeds can achieve satisfactory temporal patterns as long as enough seeds are provided.

The Effectiveness of Pattern Scoring Features. We then evaluated the performance of each pattern scoring feature: $w_1 = w_s(v, e_1, e_2)$, $w_2 = w_s(v, t_v)$, and $w_3 = w_s(t_v, t_s, t_e)$. Since our algorithm is quite accurate when all the 20 seeds are utilised, we only report pattern accuracy with three randomly selected seeds in Table 3 to make the difference more observable. We can conclude from

Table 3. The effectiveness of pattern scoring features ($|s| = 3$)

Features	nDCG$_1$	nDCG$_3$	nDCG$_5$	nDCG$_{10}$
w_1	0.58	0.52	0.51	0.48
w_2	0.57	0.52	0.50	0.48
w_3	0.57	0.55	0.53	0.49
w_1, w_2, w_3	0.67	0.59	0.57	0.52

Table 3 that all the three features are quite effective in scoring and ranking extracted temporal patterns (i.e., $nDCG_{10} \approx 0.5$). Moreover, the highest accuracy is always achieved when these three scoring features are correctly combined, which means these features can complement each other in pattern scoring under different situations.

The Effectiveness of Pattern Clustering for Score Adjustment. Considering that some temporal patterns (e.g., "get married" and "hold a wedding") are semantically related, in this work, we conduct a pattern clustering and adjust pattern scores in a cluster so that similar patterns can reciprocally reinforce each other. In this part, we evaluated the influence of semantics-based clustering on pattern accuracy. Table 4 reports the results. As before, we present average $nDCG_k$ over both relations and temporal status ($START$ and END). We can observe that pattern clustering can effectively improve the accuracy of extracted patterns especially their relative rankings.

Table 4. The effectiveness of pattern clustering

Approaches	nDCG$_1$	nDCG$_3$	nDCG$_5$	nDCG$_{10}$
With clustering	0.90	0.88	0.83	0.76
Without clustering	0.90	0.79	0.74	0.65

4 Related Work

Temporal Knowledge Harvesting. To our best knowledge, only a few previous work [8,12,14,15,22,25,26] has addressed the problem of extracting temporal knowledge. We classify them into two categories: coupling systems and extracting systems. Coupling systems [8,12,14,22,26] aim at detecting valid time scopes for existing relational facts. T-Yago [26] leverages regular expressions to extract temporal information from Wikipedia's Infoboxes, Categories and Lists, which limits its coverage and applicability to widely available free text. MS MLI [8] and UNED Blender [12] are trained by distant supervision and try to extract valid time for given relations by aggregating timestamps. Whereas, they mainly focus on seven predefined relations [21] and cannot be directly applied to general relations. Besides, they only adopt sentence-level analysis and ignore the utility of Web's redundancy. Extracting systems [15,25] try to harvest temporal knowledge from scratch, and they discover relations and their valid time from text corpus simultaneously. TIE [15] leverages Markov Logic Network (MLN) with transitivity rules to infer the temporal order of extracted events [23], which is not a specific temporal value. PRAVDA [25] uses textual patterns to represent the generated candidate facts, and then labels each candidate facts through a graph-based label propagation algorithm. However, the textual patterns utilised in PRAVDA cannot determine the beginning or end of a relation. Besides, these patterns are derived directly from raw text. As discussed in Sect. 1, natural languages are intrinsically ambiguous which limits the accuracy and coverage of patterns extracted from raw text.

Pattern-Based Information Extraction. Pattern-based information extraction systems has the advantage of high interpretability and easy to cope with errors [6]. However, early-stage systems, such as DIPRE [4] and Snowball [1], leverage heuristic rules to extract only certain predefined information, and meanwhile suffer from low precision and coverage. Recently, large-scale pattern-based information extraction systems, e.g. Probase [27], ReVeb [10], and Ollie [19], have been built. But they mainly focus on static relation extraction and ignore temporal variance of the extracted relations. To the best of our knowledge, this paper is the first one that targets at designing automatic approaches for temporal pattern extraction which, we believe, is extremely important for temporal knowledge harvesting.

5   Conclusion

In this paper, we study the problem of temporal pattern extraction which is an indispensable pre-step for temporal knowledge harvesting. We propose a novel framework to automatically discover temporal patterns by analysing a large-scale text corpus. Our experimental results on real world datasets verify the effectiveness of our proposals. As future work, we will evaluate our algorithms on other types of text corpus including news articles, blogs and tweets, etc. Besides, we will apply the extracted temporal patterns to harvest temporal facts and design an iterative framework to construct a large-scale temporal knowledge base.

References

1. Agichtein, E., Gravano, L.: Snowball: extracting relations from large plain-text collections. In: Proceedings of the Fifth ACM Conference on Digital Libraries, pp. 85–94. ACM (2000)
2. Auer, S., Bizer, C., Kobilarov, G., Lehmann, J., Cyganiak, R., Ives, Z.: DBpedia: a nucleus for a web of open data. In: Aberer, K., et al. (eds.) ASWC/ISWC -2007. LNCS, vol. 4825, pp. 722–735. Springer, Heidelberg (2007). https://doi.org/10.1007/978-3-540-76298-0_52

3. Bollacker, K., Evans, C., Paritosh, P., Sturge, T., Taylor, J.: Freebase: a collaboratively created graph database for structuring human knowledge. In: Proceedings of the 2008 ACM SIGMOD International Conference on Management of Data, pp. 1247–1250. ACM (2008)
4. Brin, S.: Extracting patterns and relations from the world wide web. In: Atzeni, P., Mendelzon, A., Mecca, G. (eds.) WebDB 1998. LNCS, vol. 1590, pp. 172–183. Springer, Heidelberg (1999). https://doi.org/10.1007/10704656_11
5. Campos, R., Dias, G., Jorge, A.M., Jatowt, A.: Survey of temporal information retrieval and related applications. ACM Comput. Surv. (CSUR) **47**(2), 15 (2015)
6. Chiticariu, L., Li, Y., Reiss, F.R.: Rule-based information extraction is dead! Long live rule-based information extraction systems! In: EMNLP, pp. 827–832, October 2013
7. Clark, K., Manning, C.D.: Deep reinforcement learning for mention-ranking coreference models. arXiv preprint arXiv:1609.08667 (2016)
8. Cucerzan, S., Sil, A.: The MSR systems for entity linking and temporal slot filling at TAC 2013. In: Text Analysis Conference (2013)
9. Ester, M., Kriegel, H.P., Sander, J., Xu, X., et al.: A density-based algorithm for discovering clusters in large spatial databases with noise. In: KDD, vol. 96, pp. 226–231 (1996)
10. Fader, A., Soderland, S., Etzioni, O.: Identifying relations for open information extraction. In: Proceedings of the Conference on Empirical Methods in Natural Language Processing, pp. 1535–1545. Association for Computational Linguistics (2011)
11. Ferragina, P., Scaiella, U.: TAGME: on-the-fly annotation of short text fragments (by Wikipedia entities). In: Proceedings of the 19th ACM International Conference on Information and Knowledge Management, pp. 1625–1628. ACM (2010)
12. Garrido, G., Penas, A., Cabaleiro, B.: UNED slot filling and temporal slot filling systems at TAC KBP 2013: system description. In: TAC (2013)
13. Hoffart, J., Suchanek, F.M., Berberich, K., Weikum, G.: YAGO2: a spatially and temporally enhanced knowledge base from Wikipedia. Artif. Intell. **194**, 28–61 (2013)
14. Kuzey, E., Weikum, G.: Extraction of temporal facts and events from Wikipedia. In: Proceedings of the 2nd Temporal Web Analytics Workshop, pp. 25–32. ACM (2012)
15. Ling, X., Weld, D.S.: Temporal information extraction. In: AAAI, vol. 10, pp. 1385–1390 (2010)
16. Mahdisoltani, F., Biega, J., Suchanek, F.: YAGO3: a knowledge base from multilingual Wikipedias. In: 7th Biennial Conference on Innovative Data Systems Research. CIDR Conference (2014)
17. Mikolov, T., Yih, W.t., Zweig, G.: Linguistic regularities in continuous space word representations. In: Proceedings of NAACL-HLT, pp. 746–751 (2013)
18. Mitchell, T., et al.: Never-ending learning (2015)
19. Schmitz, M., Bart, R., Soderland, S., Etzioni, O., et al.: Open language learning for information extraction. In: Proceedings of the 2012 Joint Conference on Empirical Methods in Natural Language Processing and Computational Natural Language Learning, pp. 523–534. Association for Computational Linguistics (2012)
20. Strötgen, J., Gertz, M.: HeidelTime: high quality rule-based extraction and normalization of temporal expressions. In: Proceedings of the 5th International Workshop on Semantic Evaluation, pp. 321–324. Association for Computational Linguistics (2010)

21. Surdeanu, M.: Overview of the TAC2013 knowledge base population evaluation: English slot filling and temporal slot filling. In: Proceedings of the Sixth Text Analysis Conference (TAC 2013) (2013)
22. Talukdar, P.P., Wijaya, D., Mitchell, T.: Coupled temporal scoping of relational facts. In: Proceedings of the Fifth ACM International Conference on Web Search and Data Mining, pp. 73–82. ACM (2012)
23. UzZaman, N., Llorens, H., Derczynski, L., Verhagen, M., Allen, J., Pustejovsky, J.: SemEval-2013 task 1: TempEval-3: evaluating time expressions, events, and temporal relations
24. Wang, Y., Dylla, M., Spaniol, M., Weikum, G.: Coupling label propagation and constraints for temporal fact extraction. In: Proceedings of the 50th Annual Meeting of the Association for Computational Linguistics: Short Papers-Volume 2, pp. 233–237. Association for Computational Linguistics (2012)
25. Wang, Y., Yang, B., Qu, L., Spaniol, M., Weikum, G.: Harvesting facts from textual web sources by constrained label propagation. In: Proceedings of the 20th ACM International Conference on Information and Knowledge Management, pp. 837–846. ACM (2011)
26. Wang, Y., Zhu, M., Qu, L., Spaniol, M., Weikum, G.: Timely YAGO: harvesting, querying, and visualizing temporal knowledge from Wikipedia. In: Proceedings of the 13th International Conference on Extending Database Technology, pp. 697–700. ACM (2010)
27. Wu, W., Li, H., Wang, H., Zhu, K.Q.: Probase: a probabilistic taxonomy for text understanding. In: Proceedings of the 2012 ACM SIGMOD International Conference on Management of Data, pp. 481–492. ACM (2012)
28. Yates, A., Cafarella, M., Banko, M., Etzioni, O., Broadhead, M., Soderland, S.: TextRunner: open information extraction on the web. In: Proceedings of Human Language Technologies: The Annual Conference of the North American Chapter of the Association for Computational Linguistics: Demonstrations, pp. 25–26. Association for Computational Linguistics (2007)

Simple SQL Validation of Generalized Entity Integrity

Zhuoxing Zhang[1], Hong Zhang[1(✉)], and Sebastian Link[2]

[1] College of Computer and Information Science, Southwest University,
Chongqing, China
zhang1010@email.swu.edu.cn, zhangh@swu.edu.cn
[2] Department of Computer Science, University of Auckland, Auckland, New Zealand
s.link@auckland.ac.nz

Abstract. Codd's rule of entity integrity stipulates the existence of a primary key over every database table. That is, uniqueness and absence of null markers are enforced on the columns of the primary key. Key sets stipulate a generalized entity integrity rule that can be achieved on data sets where primary keys do not exist. Indeed, a key set means that different pairs of rows can be distinguished by unique non-null values on potentially different elements of the key set. While primary keys are a core feature of SQL databases, key sets have not been researched much at all. Our goal is to motivate the actual use of key sets in database systems. The use of key sets depends at least on the ability to identify those key sets that are meaningful in a given application domain, and to efficiently validate such key sets during the lifetime of the database. For this purpose, we analyze for the first time the performance of validating key sets in SQL experimentally, and also conduct experiments that provide insight on the time and size required to generate Armstrong relations for the implication of unary key sets by arbitrary key sets. Armstrong relations provide computational support for identifying key sets that are meaningful for a given application domain.

1 Introduction

Keys provide efficient access to data in database systems. They are required to understand the structure and semantics of data. For a given collection of entities, a key refers to a set of column names whose values uniquely identify an entity in the collection. For example, a key for a relational table is a set of columns such that no two different rows have matching values in each of the key columns. Keys are fundamental for most data models, including semantic models, object models, XML, RDF, and graphs. They advance many classical areas of data management such as data modeling, database design, and query optimization. Knowledge about keys empowers us to (1) uniquely reference entities, (2) reduce data redundancy, (3) improve selectivity estimates in query processing, (4) feed new access paths to query optimizers, (5) access data more efficiently via physical optimization, and (6) gain new insight into application data. In modern

© Springer Nature Switzerland AG 2019
L. Chang et al. (Eds.): ADC 2019, LNCS 11393, pp. 31–44, 2019.
https://doi.org/10.1007/978-3-030-12079-5_3

applications, keys facilitate data integration, help detect anomalies, guide data repairs, and return consistent answers to queries over dirty data.

For real-life applications, data models need to accommodate missing information. SQL permits occurrences of a null marker to model any kind of missing value. Occurrences of the null marker mean that no information is available about the value of that row on that attribute. Codd's rule of entity integrity says that every entity is uniquely identifiable. SQL supports entity integrity by primary keys. A primary key is a collection of attributes which stipulates uniqueness and completeness. That is, no row of a relation must have an occurrence of the null marker on any column of the primary key, and the combination of values on the columns of the primary key must be unique.

Example 1. Consider the following snapshots of data from different wards at a hospital. The first ward tracks information about the *name* of a patient, who was treated for an *injury* in some *room* at some *time*.

Room	Name	Injury	Time
1	Miller	Cardiac infarct	Sunday, 19
⊥	⊥	Skull fracture	Monday, 19

An example of a primary key that is satisfied by this table is $\mathcal{K}_1 = \{injury, time\}$. In addition to the previous attributes, the second ward also tracks information about the *address* of patients.

Room	Name	Address	Injury	Time
2	Maier	Dresden	Leg fracture	Sunday, 16
1	Miller	Pirna	Leg fracture	Sunday, 16

An example of a primary key that is satisfied by the second table is $\mathcal{K}_2 = \{room, time\}$. Note that $\mathcal{K}_2$ is not satisfied by the first table because column *room* features a null marker occurrence, while the second table does not satisfy $\mathcal{K}_1$ because both rows have matching non-null values on *injury* and *time*.

In practice, requiring a primary key over every database table is often inconvenient or not achievable. This is particularly true for modern applications such as data integration and big data. Indeed, it can happen easily that a given relation does not exhibit any primary key. This is illustrated by the following example.

Example 2. We continue our previous example by looking at a data snapshot that results from integrating the previous two data sets from Example 1. Using null markers in the column *address* for the rows of the first table, the result of the data integration process brings forward the following snapshot.

Room	Name	Address	Injury	Time
1	Miller	⊥	Cardiac infarct	Sunday, 19
⊥	⊥	⊥	Skull fracture	Monday, 19
2	Maier	Dresden	Leg fracture	Sunday, 16
1	Miller	Pirna	Leg fracture	Sunday, 16

Evidently, the snapshot does neither satisfy the primary key $\mathcal{K}_1$ nor the primary key $\mathcal{K}_2$. In fact, the snapshot does not satisfy any primary key since each column features some null marker occurrence, or a duplication of some value.

In response, several researchers proposed the notion of a key set. As the term suggests, a key set is a set of attribute subsets. Naturally, we call the elements of a key set a key. A relation satisfies a given key set if for every pair of distinct rows in the relation there is some key in the key set on which both rows have no null marker occurrences and non-matching values on some attribute of the key. The flexibility of a key set over a primary key can easily be recognized, as a primary key would be equivalent to a singleton key set, with the only element being the primary key. Indeed, with a key set different pairs of rows in a relation may be distinguishable by different keys of the key set, while all pairs of rows in a relation can only be distinguished by the same primary key. We illustrate the notion of a key set on our running example.

Example 3. The relation in Example 2 satisfies no primary key. Nevertheless, the relation satisfies several key sets. For example, the key set $\{\{room\}, \{time\}\}$ is satisfied, but not the key set $\{\{room, time\}\}$. The relation also satisfies the key sets $\mathcal{X}_1 = \{\{room, time\}, \{injury, time\}$ and $\mathcal{X}_2 = \{\{name, time\}, \{injury, time\}\}$.

Both primary keys and key sets are independent of the interpretation of null marker occurrences. That is, any given primary key and any given key set is either satisfied or not, independently of what information any of the null marker occurrences represent. The importance of this independence is particularly appealing in modern applications where data is integrated from various sources, and different interpretations may be required for different occurrences of null markers.

Given the flexibility of key sets over primary keys, it seems natural to further investigate the notion of a key set. Neither the research community nor any system implementations have analyzed key sets since their proposal in 1989. Very recently, Hannula and Link [5] did study the implication problem associated with key sets. They established automated reasoning capabilities for keys sets that facilitate the processing of database queries and updates. However, the usability of key sets by industry would require even more basic capabilities. Two of these capabilities are *acquisition* and *validation*.

Acquisition refers to the ability of business analysts to identify key sets that model business rules in an application domain. It has been shown [10] that Armstrong relations provide computational support for business analysts to communicate their understanding of the application domain to domain experts. An inspection of the Armstrong relations can provide valuable feedback to the analysts in the form of any business rules that they were not able to identify before. While [5] has brought forward an algorithm that computes an Armstrong relation for the implication of unary by arbitrary key sets, the actual size of these Armstrong relations and the time to compute them has not been analyzed. For the *acquisition* of key sets we need to know how long it takes to compute Armstrong relations and how many rows they carry. If it takes too long or the Armstrong relations are too large, the computational support for an effective dialogue between business analysts and domain experts may be infeasible. Furthermore, *validation* refers to the problem of validating whether a given key set is satisfied by a given table. Without efficient means to validate a collection of key sets it is impossible to utilize them for the purpose of data management. For the industry standard, it is important to understand the performance of validating key sets within SQL systems.

Contributions. We address the performance of acquiring and validating key sets. (1) We compare key sets with other uniqueness constraints in databases. (2) We show how the validation of key sets can be done in SQL. Indeed, we analyze to which degree a given key set is valid, for example, the percentage of rows in a table that do not contribute to its violation. (3) We analyze the time to execute the validation queries on real-world data sets with missing information. The first takeaway is that the validation time grows quadratically in the underlying number of rows. The second takeaway is that for key sets with the same number of attributes, those with a larger number of elements result in fewer violations and faster validation. (4) We analyze the time to compute Armstrong relations for the implication of unary by arbitrary key sets. As the computation is precisely exponential in the given key sets, we conduct experiments to determine the average size and computation time.

Organization. We discuss related work in Sect. 2. Basic notions and notation are fixed in Sect. 3. The validation problem is analyzed in Sect. 4. The computation of Armstrong relations and their experimental evaluation is described in Sect. 5. Section 6 contains a conclusion and outlook to future work.

2   Related Work

Codd proposed the rule of entity integrity, which stipulates that every entity in every table should be uniquely identifiable. In SQL that led to primary keys, which are distinguished candidate keys. An attribute set is a *candidate key* for a given relation if every pair of distinct tuples has no null marker occurrences on any of the attributes of the candidate key and non-matching values on some attribute of the candidate key [12]. Candidate keys are singleton key sets, that is, key sets with one element [6]. As the relation from Example 2 shows, there are relations on which no candidate key holds, but which satisfy key sets.

Lucchesi and Osborn studied computational problems of candidate keys [12]. They proved that deciding whether a given relation satisfies some key of cardinality not greater than some given positive integer is NP-complete. Recently, this problem was shown to be W[2]-complete in the key [1]. The discovery which key sets hold on a given relation is beyond our scope.

Key sets are a generalization of Codd's rule for entity integrity [14]. Extremal problems of unary key sets were studied in [13]. Key sets were further discussed in [11] where also Codd's rule for referential integrity was generalized. Recently, a binary axiomatization, the coNP-completeness of the implication problem, and the non-existence of Armstrong relations for arbitrary key sets were established [5], but also how to compute them for unary by arbitrary key sets. We continue this inquiry by asking how feasible the computation of Armstrong relations is for the acquisition problem. The validation problem of key sets has also not been considered previously.

Possible and certain keys were also proposed recently [9]. Certain keys correspond to key sets which have only singleton keys as elements. The paper [9] investigate computational problems for possible and certain keys.

Contextual keys separate completeness from uniqueness requirements [15]. They are expressions (C, X) where $X \subseteq C$, and different from key sets since $X \subseteq C$ is a key for only those tuples that are complete on C. In particular, the case where $C = X$ only requires uniqueness on X for tuples that are complete on X. This captures the SQL UNIQUE constraint.

Keys have also been investigated in XML [7], graphs [4], and uncertain data [2,8].

3 Preliminaries

We give some basic definitions and fix notation. A *relation schema* is a finite non-empty set of attributes, usually denoted by R. A *relation* r over R consists of tuples t that map each $A \in R$ to $dom(A) \cup \{\bot\}$ where $dom(A)$ is the domain associated with attribute A and $\bot$ is the unique null marker. Given a subset X of R, we say that a tuple t is X-*total* if $t(A) \neq \bot$ for all $A \in X$. Moreover, $dom(A)$ represents the possible values that can occur in column A of a table, and $\bot$ represents missing information. That is, if $t(A) = \bot$, then there is no information about the value $t(A)$ of tuple t on attribute A.

We will use WARD = {*room, name, address, injury, time*} as the relation schema of a running example. Each row of the table in Example 2 represents a tuple. The second row is {*injury, time*}-total, but not total on any proper superset of {*injury, time*}. The four tuples together constitute a relation over WARD. The following definition introduces the central object of our studies. It was first defined by Thalheim [14].

Definition 1. *A key set is a finite, non-empty collection $\mathcal{X}$ of subsets of a given relation schema R. We say that a relation r over R satisfies the key set $\mathcal{X}$ if and only if for all distinct $t, t' \in r$ there is some $X \in \mathcal{X}$ such that t and t' are X-total and $t(X) \neq t'(X)$. A key set that consists of only singletons is called a* unary key set.

We write $\mathcal{X}, \mathcal{Y}, \mathcal{Z}, \ldots$ for key sets, $X, Y, Z, \ldots$ for attribute sets, and $A, B, C, \ldots$ for attributes. We use A to denote the singleton $\{A\}$. The *cardinality* of $\mathcal{X}$ is the total number of attribute occurrences, $\|\mathcal{X}\| = \sum_{\mathcal{K} \in \mathcal{X}} |\mathcal{K}|$, and the *size* of $\mathcal{X}$ is the number $|\mathcal{X}|$ of its elements. For example, the key set $\{\{injury, time\}, \{room, name, time\}\}$ has cardinality five and size two.

4   Validation in SQL

So far, research on key sets has only been theoretical [5,11,14]. While the idea is natural, it will only be useful when implemented in database systems. A fundamental question for the uptake of key sets by SQL is how to validate them on a given table. The first aim is to identify a simple SQL query that does this job. We will identify a query that returns all the rows of a given table that participate in a violation of the given key set. The second aim is to evaluate the performance of this query.

4.1   SQL Queries for Key Set Validation

We aim at finding an SQL query that allows us to validate whether a given key set

$$\mathcal{X} = \{\{A_1^1, \ldots, A_{n_1}^1\}, \ldots, \{A_1^k, \ldots, A_{n_k}^k\}\}$$

is satisfied by a given table.

According to Definition 1 for r to satisfy $\mathcal{X}$, for each pair of distinct tuples $t, t' \in r$ there must be some $\mathcal{K}_i = \{A_1^i, \ldots, A_{n_i}^i\} \in \mathcal{X}$ such that both t and t' are $\mathcal{K}_i$-total and for some $A \in \mathcal{K}_i$, $t(A) \neq t'(A)$. Unfortunately, SQL does not have a simple way of expressing universal quantification, so we will need to reformulate this query in terms of a negated existential quantification. Indeed, for r to satisfy $\mathcal{X}$ there must not be a tuple $t \in r$ such that there is a tuple $t' \in r$ distinct from t such that for all $\mathcal{K}_i = \{A_1^i, \ldots, A_{n_i}^i\} \in \mathcal{X}$, t or t' are not $\mathcal{K}_i$-total, and for all $A \in \mathcal{K}_i$, $t(A) = t'(A)$. In fact, instead of saying that there must not be such a tuple $t \in r$, we could let our query return all such tuples $t \in r$. If the result is empty, then the key set is valid, otherwise we have collected all the tuples that participate in a violation of the key set. The latter would give us a more detailed analysis since it informs us about the degree by which the given key set is valid in the given table.

Before we stipulate the SQL query, we need to accommodate the fact that the given table may contain duplicate tuples. Since we want to handle this situation, we assume that there is a surrogate key column called *id* which associates a unique identifier with any row in the table. Note that this is what all SQL-based systems do internally anyway.

$\mathcal{Q}_1$:
```
SELECT DISTINCT R.*
FROM    R, R AS R'
WHERE   (R.id <> R'.id) AND
        (R.A₁¹ IS NULL OR ... OR R'.Aₙ₁¹ IS NULL OR
```

$\mathcal{Q}_1$:

SELECT DISTINCT $R.*$

FROM $\quad R, R$ AS R'

WHERE $\quad (R.id <> R'.id)$ AND

$\quad (R.A_1^1$ IS NULL OR $\ldots$ OR $R'.A_{n_1}^1$ IS NULL OR

$\quad\quad (R.A_1^1 = R'.A_1^1$ AND $\ldots$ AND $R.A_{n_1}^1 = R'.A_{n_1}^1))$ AND

$\ldots$

$\quad (R.A_1^k$ IS NULL OR $\ldots$ OR $R'.A_{n_1}^k$ IS NULL OR

$\quad\quad (R.A_1^k = R'.A_1^k$ AND $\ldots$ AND $R.A_{n_k}^k = R'.A_{n_k}^k))$;

Note that the self-join in the FROM clause of $\mathcal{Q}_1$ reflects the fact that all tuple pairs need to be compared. This is a consequence of wanting to distinguish all pairs of distinct tuples in the relation by some element of $\mathcal{X}$. This also indicates that the growth of the time required to evaluate this query on the given table will grow quadratically in the size of the table. In the scope of this paper we do not aim at investigating how this performance can be improved by logical or physical optimization techniques.

Also note that we can simply add a LIMIT 1 clause to the end of the query $\mathcal{Q}_1$ in case we only want to validate whether the given key set is satisfied by the given table. In this case, the query will terminate as soon as it finds one tuple that participates in the violation of the key set.

Our query $\mathcal{Q}_1$ answers the question *Which tuples cannot be distinguished from all of the other tuples by the given key set?* Consequently, the calculation

$$Valid\ tuples = |r| - |\mathcal{Q}_1|$$

gives us the number of tuples in the given table which can be distinguished from all of the other tuples by the given key set, and

$$Fraction\ of\ valid\ tuples = \frac{|r| - |\mathcal{Q}_1|}{|r|}$$

would give us the fraction of tuples in the given table that can be distinguished from all of the other tuples by the given key set. These constitute one solution to the degree of validity by which a given key set holds on a given table.

A different solution would quantify the degree of validity by the number/fraction of tuple pairs in the given table that can be distinguished by the given key set. For this, we can use the query $\mathcal{Q}_2$ which results from $\mathcal{Q}_1$ by replacing SELECT DISTINCT $R.*$ by SELECT $*$. Indeed, $\mathcal{Q}_2$ would return all those pairs of distinct tuples which cannot be distinguished by the given key set. Since the number of pairs of distinct tuples in a given table r is $\binom{r}{2} = 1/2 \times (|r|^2 - |r|)$, and $\mathcal{Q}_2$ compares each pair of tuples twice, the calculation

$$Valid\ tuple\ pairs = 1/2 \times (|r|^2 - |r|) - 1/2 \times |\mathcal{Q}_2|$$

gives us the number of tuple pairs in the given table which can be distinguished by the given key set, and

$$Fraction\ of\ valid\ tuple\ pairs = \frac{1/2 \times (|r|^2 - |r|) - 1/2 \times |\mathcal{Q}_2|}{1/2 \times (|r|^2 - |r|)}$$

would give us the fraction of tuple pairs in the given table that can be distinguished by the given key set.

Example 4. Recall our data set from Example 2, which we augment now by the surrogate key column *id* as follows.

Id	Room	Name	Address	Injury	Time
1	1	Miller	⊥	Cardiac infarct	Sunday, 19
2	⊥	⊥	⊥	Skull fracture	Monday, 19
3	2	Maier	Dresden	Leg fracture	Sunday, 16
4	1	Miller	Pirna	Leg fracture	Sunday, 16

A computation of the various measures is then applied to the key sets $\{\mathcal{K}_1\}$ and $\{\mathcal{K}_2\}$ from Example 3, and the key set $\mathcal{X} = \{\mathcal{K}_1, \mathcal{K}_2\}$ with respect to the given table. The results are as follows.

Measure\Key set	$\{\mathcal{K}_2\}$	$\{\mathcal{K}_1\}$	$\mathcal{X}$
Valid tuples	0	2	4
Fraction of valid tuples	0	1/2	1
Valid tuple pairs	3	5	6
Fraction of valid tuple pairs	1/2	5/6	1

For example, no tuples in the given table is distinguishable from every other tuple by $\{\mathcal{K}_2\}$, while the first two tuples are each distinguishable from all the other tuples by $\{\mathcal{K}_1\}$, and every tuple is distinguishable from all the other tuples by $\mathcal{X}$. Similarly, tuple pairs $(1,3)$, $(1,4)$, and $(3,4)$ can be distinguished by $\{\mathcal{K}_2\}$, while only tuple pair $(3,4)$ cannot be distinguished by $\{\mathcal{K}_1\}$, and every tuple pair can be distinguished by $\mathcal{X}$.

4.2   Experiments

The previous section has provided some answers on how to validate given key sets on given data sets within an SQL-based database management system. This section will give some first insight into how long the validation of key sets may take in practice, how the time for validation grows in terms of the input size, and how the degree of validity is influenced by the characteristics of the key sets. In what follows we will first describe the real-world data sets we are using for our experiments. We will then analyse the validation time, followed by an analysis of the validity degrees. Our experiments are conducted on an Intel Core i7-7500U CPU machine, which has 8 GB RAM and 2.7 GHz.

Data Sets. Our experiments are conducted on real-world data sets with incomplete information. The data sets comprise of hepatitis, horse, plista, flight, ncvoter, and uniprot, which have been used as benchmark data sets for estimating the performance of algorithms that discover different kinds of data dependencies from data. They are publicly available[1]. The major characteristics of these

Table 1. Characteristics of real-world incomplete data sets

Name	#columns	#rows	#null
Hepatitis	20	155	167
Horse	28	300	1,605
Plista	63	1,000	23,317
Flight	109	1,000	51,938
Ncvoter	19	32,000	97,924
Uniprot	30	32,000	233,583

data sets are detailed in Table 1. In the table, *name* denotes the unique name of the data set, *#columns* denotes the number of columns, *#rows* denotes the number of rows, and *#null* denotes the number of null marker occurrences in the data set.

Validation Time. Our first experiment concerns the time of evaluating query Q_1 for randomly created key sets of varying size and varying fragments of the given data sets. For each data set, we created six fragments with growing numbers of rows. Similarly, for each data set we randomly created key sets of varying size but where the cardinality is fixed to the number of columns in the data set. For each key set size, we randomly created ten key sets of that size, and then measured the time to evaluate Q_1 on each of the fragments and each of the key sets. The output (y-axis) is the average time of evaluating Q_1 for the ten key sets and for the given fragment (x-axis). Figure 1 shows the results on (the fragments of) our data sets and for various key set sizes.

There are mostly three main observations. (1) Validation times are very fast for modest numbers of rows, even for scanning all tuple pairs. For example, the longest time was spent on the full data set *flight* for validating key sets of size 1 and cardinality 63. This took 30 s, but most other times are significantly smaller, in particular for key sets of larger size. The two larger data sets with 32,000 rows exemplify that the quadratic time complexity has its consequences: validation on full ncvoter takes just under 3hrs and validation on full uniprot takes about 5 and a half hours. (2) The validation time is quadratic in the size of the data sets, for each fixed key set size. This is not surprising as we need to compare all tuple pairs. (3) Among key sets of the same cardinality, the larger the size of a key set, the faster the validation is. This is a consequence of the observation that more violating tuples are found for smaller key set sizes, see the next experiment. Hence, more time needs to be spent on identifying those.

---

[1] https://hpi.de/naumann/projects/repeatability/data-profiling/fds.html.

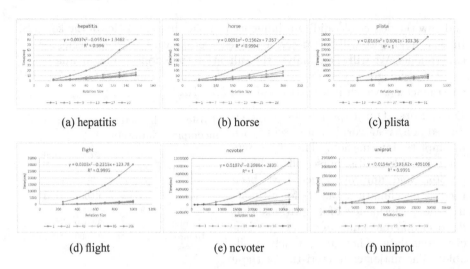

Fig. 1. Average validation times for growing sizes of data sets and key sets

Validity Degrees. Figure 2 shows the number of tuples in the output of query Q_1, when evaluated with respect to the given key set and the given data set fragment. There are two main observations. (1) The number of violations, and thus the degree of violation, grows linearly in the size of the data set, for each given key set size. (2) Among the key sets of the same cardinality, the larger the size of a key set, the larger the degree of validity. Indeed, if the key set sizes are smaller, then tuple pairs become harder to distinguish (since matching non-null values are required on more attributes).

Final Remarks. We conclude that key set validation in SQL-based systems is very much feasible, and can provide interesting insight into the compliance of the data set with generalized entity integrity. Among key sets of the same cardinality, those of smaller size might be favorable since they can distinguish more entities and are faster to validate.

5   Acquisition of Key Sets by Armstrong Relations

We examine the feasibility of computational support for the acquisition of key sets from a performance point of view. Armstrong relations have been found useful for the acquisition of meaningful business rules [10]. Intuitively, this is no different for keys sets. Recently, it was shown that Armstrong relations do not exist for all collections of key sets, but an algorithm has been established that computes an Armstrong relation for the implication of unary key sets by any given collection of arbitrary key sets. Here, we will provide some first experimental evidence that the size of these Armstrong relations and the time to compute them are small enough to offer computational support in practice. For that purpose we briefly recall the computation of Armstrong relation for key sets.

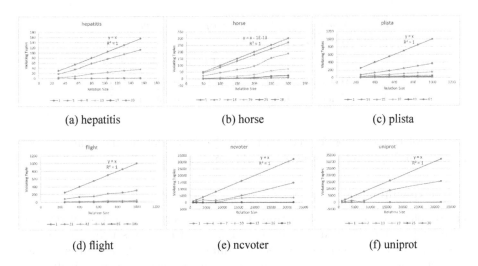

(a) hepatitis   (b) horse   (c) plista

(d) flight   (e) ncvoter   (f) uniprot

Fig. 2. Number of violating tuples for growing sizes of data sets and key sets

5.1   Computation of Armstrong Relations

The implication problem of unary key sets $\mathcal{X}$ by $\Sigma = \{\mathcal{X}_1, \ldots, \mathcal{X}_n\}$ of arbitrary key sets only depends on the attributes in each given key set of Σ, and not on how they are grouped as sets in a key set [5]. We thus identify $\mathcal{X}$ with $\bigcup \mathcal{X}$ and each $\mathcal{X}_i$ with $\bigcup \mathcal{X}_i$. We then compute *anti-keys*, the maximal key sets not implied by Σ. Given the anti-keys, an Armstrong relation for Σ is generated by starting with a single complete tuple, and introducing for each anti-key a new tuple that has matching total values on the attributes of the anti-key and unique values elsewhere. This construction ensures that all non-implied (unary) key sets are violated and all given key sets are satisfied. The computation of the anti-keys from Σ is done by taking the complements of the minimum transversals of the hypergraph formed by the elements of Σ. A transversal for a given set of attribute subsets $\mathcal{X}_i$ is an attribute subset $\mathcal{T}$ such that $\mathcal{T} \cap \mathcal{X}_i \neq \emptyset$ holds for all i. While many efficient algorithms exist for the computation of all hypergraph transversals, it is still an open problem whether there is an algorithm that is polynomial in the output [3]. This construction always generates an Armstrong relation whose number of tuples is at most quadratic in a minimum-sized Armstrong relation [5].

Example 5. Consider the set $\Sigma = \{\mathcal{X}_1, \mathcal{X}_2\}$ with $\mathcal{X}_1$ and $\mathcal{X}_2$ from Example 3. Then $\bigcup \mathcal{X}_1 = \{room, time, injury\}$ and $\bigcup \mathcal{X}_2 = \{name, time, injury\}$. The minimum transversals are $\mathcal{T}_1 = \{time\}$, $\mathcal{T}_2 = \{injury\}$, and $\mathcal{T}_3 = \{room, name\}$, and their complements are anti-keys $\mathcal{A}_1 = \{room, name, address, injury\}$, $\mathcal{A}_2 = \{room, name, address, time\}$, and $\mathcal{A}_3 = \{address, injury, time\}$. The following relation is Armstrong for Σ.

Room	Name	Address	Injury	Time
1	Miller	24 Queen St	Leg fracture	Sunday, 16
1	Miller	24 Queen St	Leg fracture	Monday, 19
1	Miller	24 Queen St	Arm fracture	Monday, 19
2	Maier	24 Queen St	Arm fracture	Monday, 19

The relation satisfies $\mathcal{X}_1$ and $\mathcal{X}_2$, but the relation violates the unary key set $\varphi' = \{\{room\}, \{name\}, \{address\}, \{time\}\}$, so φ' is not implied by Σ.

5.2 Experiments

The problem of computing Armstrong relations for a given collection of key sets is precisely exponential in the given cardinality of the key set [5]. Hence, the question arises for which input sizes the computations are efficient in providing support to analysts and domain experts for the acquisition of meaningful key sets. For this purpose we conduct experiments in which we apply the computation to randomly created unary key sets over every schema with a fixed number $|R|$ of attributes, with $|R|$ varying from 2 to 15. For each $|R|$ (x-axis), and each cardinality $n = 2, \ldots, |R|$ we randomly generate ten key sets of cardinality n, and then compute the average number of tuples in the output (y-axis) and the average time to compute the output (y-axis), respectively. The results are illustrated in Fig. 3 (a) and (b), respectively.

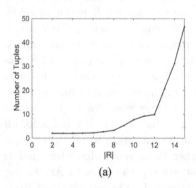

(a)

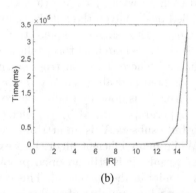

(b)

Fig. 3. Number of tuples (a) and time (b) for computing Armstrong relations for randomly created key sets over relation schemata R of various size $|R|$

For schemata with up to 12 attributes, Armstrong relations with an average of up to ten tuples are created, and with average time of up to ten seconds. It is therefore reasonable to say that the computation of Armstrong relations is feasible and can offer computational support for the acquisition problem. Even for schemata with 14 attributes, the average size is 30 tuples and average time 50 s to compute.

6   Conclusion and Future Work

Key sets can distinguish entities that primary keys cannot. For their uptake, effective solutions for their validation and acquisition problems are required. We have proposed first SQL queries to address validation, and studied their performance on real-world benchmarks. Validation is quadratic in the number of rows, and more effective and efficient when the elements of a key set have fewer attributes. We have also tested the average performance of computing Armstrong relations for key sets. For up to 12 attributes the computation is fast and creates outputs of small sizes. This confirms the feasibility for computational support of the acquisition problem. Dedicated algorithms instead of SQL-based solutions to the validation problem would avoid disk-based access. It is also important to study the impact of index structures on validation times. Our algorithms for computing Armstrong relations can uncover insight on how these relations support acquisition.

References

1. Bläsius, T., Friedrich, T., Schirneck, M.: The parameterized complexity of dependency detection in relational databases. In: Guo, J., Hermelin, D. (eds.) 11th International Symposium on Parameterized and Exact Computation, IPEC 2016, August 24-26, 2016, Aarhus, Denmark. LIPIcs, vol. 63, pp. 6:1–6:13. Schloss Dagstuhl - Leibniz-Zentrum fuer Informatik (2016)
2. Brown, P., Link, S.: Probabilistic keys. IEEE Trans. Knowl. Data Eng. **29**(3), 670–682 (2017)
3. Eiter, T., Gottlob, G., Makino, K.: New results on monotone dualization and generating hypergraph transversals. SIAM J. Comput. **32**(2), 514–537 (2003)
4. Fan, W., Fan, Z., Tian, C., Dong, X.L.: Keys for graphs. PVLDB **8**(12), 1590–1601 (2015)
5. Hannula, M., Link, S.: Automated reasoning about key sets. In: Galmiche, D., Schulz, S., Sebastiani, R. (eds.) IJCAR 2018. LNCS (LNAI), vol. 10900, pp. 47–63. Springer, Cham (2018). https://doi.org/10.1007/978-3-319-94205-6_4
6. Hartmann, S., Leck, U., Link, S.: On Codd families of keys over incomplete relations. Comput. J. **54**(7), 1166–1180 (2011)
7. Hartmann, S., Link, S.: Efficient reasoning about a robust XML key fragment. ACM Trans. Database Syst. **34**(2), 10 (2009)
8. Koehler, H., Leck, U., Link, S., Prade, H.: Logical foundations of possibilistic keys. In: Fermé, E., Leite, J. (eds.) JELIA 2014. LNCS (LNAI), vol. 8761, pp. 181–195. Springer, Cham (2014). https://doi.org/10.1007/978-3-319-11558-0_13
9. Köhler, H., Leck, U., Link, S., Zhou, X.: Possible and certain keys for SQL. VLDB J. **25**(4), 571–596 (2016)
10. Langeveldt, W., Link, S.: Empirical evidence for the usefulness of Armstrong relations in the acquisition of meaningful functional dependencies. Inf. Syst. **35**(3), 352–374 (2010)
11. Levene, M., Loizou, G.: A generalisation of entity and referential integrity in relational databases. ITA **35**(2), 113–127 (2001)
12. Lucchesi, C.L., Osborn, S.L.: Candidate keys for relations. J. Comput. Syst. Sci. **17**(2), 270–279 (1978)

13. Thalheim, B.: Dependencies in Relational Databases. Teubner, Stuttgart (1991)
14. Thalheim, B.: On semantic issues connected with keys in relational databases permitting null values. Elektronische Informationsverarbeitung und Kybernetik **25**(1/2), 11–20 (1989)
15. Wei, Z., Link, S., Liu, J.: Contextual keys. In: Mayr, H.C., Guizzardi, G., Ma, H., Pastor, O. (eds.) ER 2017. LNCS, vol. 10650, pp. 266–279. Springer, Cham (2017). https://doi.org/10.1007/978-3-319-69904-2_22

A Versatile Framework for Painless Benchmarking of Database Management Systems

Lexi Brent and Alan Fekete[✉]

The University of Sydney, Sydney, NSW 2006, Australia
{lexi.brent,alan.fekete}@sydney.edu.au

Abstract. Benchmarking is a crucial aspect of evaluating database management systems. Researchers, developers, and users utilise industry-standard benchmarks to assist with their research, development, or purchase decisions, respectively. Despite this ubiquity, benchmarking is usually a difficult process involving laborious tasks such as writing and debugging custom testbed scripts, or extracting and transforming output into useful formats. To date, there are only a limited number of comprehensive benchmarking frameworks designed to tackle these usability and efficiency challenges directly.

In this paper we propose a new versatile benchmarking framework. Our design, not yet implemented, is based on exploration of the benchmarking practices of individuals in the database community. Through user interviews, we identify major pain points these people encountered during benchmarking, and map these onto a pipeline of processes representative of a typical benchmarking workflow. We explain how our proposed framework would target each process in this pipeline, potentiating significant overall usability and efficiency improvements. We also contrast the responses of engineers working in industry with those of researchers, and examine how database benchmarking requirements differ between these two groups. The framework we propose is based around traditional synthetic workloads, would be simple to configure, highly extensible, could support any benchmark, and write output to any well-defined data format. It would collect and track all generated events, data, and relationships from the benchmark and underlying systems, and offer simple reproducibility. Complex scenarios such as distributed-client and multi-tenant benchmarks would be simplified by the framework's workload partitioning, client coordination, and output collation capabilities.

Keywords: Benchmark · TPCC · YCSB · DBMS ·
Performance evaluation

1 Introduction

Benchmarking database management systems (DBMSes) is critical for evaluating their correctness, performance, and efficacy. Organizations often employ

© Springer Nature Switzerland AG 2019
L. Chang et al. (Eds.): ADC 2019, LNCS 11393, pp. 45–56, 2019.
https://doi.org/10.1007/978-3-030-12079-5_4

industry-standard benchmarks before making purchase decisions; researchers use benchmarks to evaluate novel technology; and developers frequently run benchmarks during development to identify bugs or bottlenecks in their systems.

Due to the large variety of different DBMSes, benchmarks, database paradigms, and data formats available today, benchmarking has become an unnecessarily complex undertaking typically involving laborious manual processes and repetitive tasks [4,5,14]. This heterogeneity led to the development of new tools assisting with aspects of benchmarking, such as data and workload generation [1,2,12,13,16], precise control over request rate and transaction mixture [18], collection of statistics and environment information, and workload- or target-specific testbeds or frameworks [8,9,19]. With the rise of cloud-based distributed DBMSes, new benchmarking frameworks such as [15] have emerged targeting properties such as horizontal scaling, elasticity, and availability, with support for automated provisioning of cloud resources.

Previous research has largely focused on distinct benchmarking sub-processes or specific scenarios, rather than taking a holistic approach. For instance, OLTP-Bench [8] provided extensible support for running industry-standard benchmarks targeting relational DBMSes with a focus on fine-grained control over request rates, transaction mixtures, and access distributions; YCSB [6] provided a benchmark for large-scale distributed cloud database systems; YCSB+T [7] extended YCSB with support for transactional workloads; the TPC [17] benchmarks focused on performance evaluation of relational DBMSes; MTCB [20] provided a benchmark for multi-tenant OLTP systems; UDBMS [11] implemented a data model for benchmarking multi-model database systems; and MUDD [16], PSDG [10], PDGF [12], and NoWog [1] provided automated test data generation.

Only a few studies [3–5] have been concerned with building a comprehensive, extensible framework focusing on usability and the whole benchmarking process. Most notably, BenchFoundry [3] implemented support for deterministic trace-based workload generation within an extensible distributed benchmarking framework capable of supporting several SQL and NoSQL systems. Deterministic, trace-based workload generation makes it difficult to control the statistical distribution of inputs to match real-world situations. Implementations of traditional benchmarks within BenchFoundry may produce results inconsistent with synthetic workloads based on random sampling from a statistical distribution [5]. Additionally, BenchFoundry does not collect detailed environment metadata from benchmark clients and servers. Such metadata is often critical for assessing the validity of performance benchmarks, which need to be conducted under tightly controlled conditions and should not be impacted by resource or benchmarking bottlenecks [5].

In practice, benchmarking platforms are typically based on a series of shell scripts and configuration files that handle everything from collecting environment information to transforming benchmark output into a useful data format, in addition to executing the benchmark. Depending on the experiment, there may be several different versions of each script or configuration file typically distinguished by "meaningful" filenames. Little imagination is required to realize this often becomes chaotic. These scripts are often developed from scratch and specialised to specific systems or benchmarks, and therefore not easily adaptable. As we show in Sect. 2, these custom scripts are a major time sink and source of bugs in benchmarking workflows.

In this paper, we propose and envision a new benchmarking framework towards solving these issues and improving practices. Unlike most previous work, our proposal has a focus on usability. Uniquely, it is based on interviews revealing the real benchmarking practices of several academics and industry professionals. The result would be a highly general, extensible, and versatile framework incorporating the whole benchmarking process; from DBMS and benchmark configuration to output processing and statistical analysis, with a focus on usability and meeting the benchmarking needs of both industry and academia. While our framework proposal incorporates some ideas previously presented in the OLTP-Bench [8] and BenchFoundry [3] papers, our focus is fundamentally at a higher level. Our vision aims to remove difficulties reported by some highly experienced people; it uses traditional synthetic workloads based on sampling from a statistical distribution, while still focusing heavily on experimental repeatability.

The main contributions of this paper are:

1. An interview-based analysis of pain points in current practices and identification of similarities and differences between academia and industry.
2. The design of a new benchmarking framework addressing the pain points we identified.

The remainder of this paper is structured as follows. In Sect. 2 we describe our exploration of the benchmarking practices of academics and industry professionals. Based on this, we identify a set of major pain points in benchmarking processes and contrast the responses of industry professionals with those of academics. We show that benchmarking may be encapsulated by a *pipeline* of key processes, and we map the major pain points onto this pipeline. In Sect. 3, we describe and envision our proposed framework alongside example use cases. Section 4 describes possible avenues for future work.

2 Current Benchmarking Practices

The design and functionality of our new benchmarking framework envisioned in Sect. 3 is heavily informed by awareness of current benchmarking practices. We interviewed five people in order to gain a deeper understanding of the spectrum of practices currently employed within the community. These interviews focused on identifying pain points, time sinks, and potential improvements within respondents' existing processes.

2.1 Interview Process

Interview questions covered three broad areas: (i) **systems** including DBMSes under test, benchmarking infrastructure, and benchmarking tools; (ii) **processes** including experiment configuration, workload, data collection, statistical analysis, transformation of raw output data, storage and management of results, and measurement dimensions; and (iii) **features/functionality** desired in a benchmarking framework. We also provided respondents with a list of potential features for our new framework, and asked which features would be most applicable to their workflow. Finally, respondents were given an opportunity to suggest new features to resolve existing issues they had identified in their own benchmarking processes. Our full set of interview questions is available online[1].

Interviewees included some academic researchers, and some engineers working in industry on deployed DBMS systems. All respondents were asked identical questions regardless of their respective backgrounds, and encouraged to provide as much detail as possible. Some additional impromptu questions were asked to clarify responses or request further detail. Interviews were conducted verbally in-person or via teleconference, in a single block of time between 30 and 45 min, with responses transcribed as the respondents spoke. While the number of people involved is small, they cover a variety of situations, and so we expect that improving the issues they mentioned will have wide benefits.

These interviews were conducted in December 2015. In March 2018 we conducted a follow-up email asking some of the original respondents from both industry and academia if any significant changes to their benchmarking tools, processes, or methodology had occurred since 2015. In their replies, they reported no significant changes. Hence, we are confident that our analysis of pain points, and our proposed framework, remain relevant to the community in 2018.

2.2   Insights of Interest

Based on interview responses, we created a summary of the key challenges faced by each respondent in their benchmarking workflows. We then used those summaries to build the following taxonomy, in which each "pain point" corresponds to a key challenge raised by at least one respondent:

PP1. Initial setup and configuration. The deployment and configuration of a benchmarking experiment is often time-consuming and unintuitive.

PP2. Script writing. It is often necessary to write and debug custom testbed scripts, which is laborious and time-consuming.

PP3. Reproducibility. Repeatability and reproducibility are difficult to accomplish, usually involving a manual process of referring to multiple information sources to configure and re-execute an experiment.

PP4. Debugging. Unexpected results are difficult to substantiate, usually requiring time-consuming manual debugging.

PP5. Distributed clients. Distributed benchmarks often require manual coordination of clients and collation of output.

PP6. Metadata collection. Collecting additional system metadata (e.g. system calls) during a benchmark run requires writing custom scripts, coordinating these, and manually correlating output.

---

[1] https://github.com/lexibrent/benchfw-resources/blob/master/interview-qns.pdf.

PP7. Log correlations. Correlating events recorded by benchmark clients with those recorded in DBMS logs requires manual inspection or custom scripts.

PP8. Statistical analysis. Statistical exploration of benchmark metrics (e.g. computing correlation coefficients) is often fruitful but typically too laborious and time-consuming to be feasible.

Some notable similarities and differences were observed between the responses from researchers and industry professionals:

- Industry respondents tended to focus more on reliability and efficiency than academic respondents. For example, industry respondents expressed a desire to measure "consistency of throughput", "response time variance", and "latency with a threshold".
- Industry respondents tended to focus on applications surrounding debugging and continuous integration, whereas academic respondents primarily focused on scientific applications such as experimentation with novel technologies.
- Academic respondents were more concerned with statistical and experimental validity and repeatability than industry respondents. For example, academic respondents discussed conducting multiple trials, and methods of dealing with statistical outliers. Industry respondents did not pay much attention to these topics, with some indicating they would typically only run a benchmark multiple times to assist with debugging, rather than to improve statistical reliability.
- All respondents indicated they use cloud services such as Amazon EC2 extensively in their benchmarking processes.
- All respondents agreed that distributed benchmark clients are difficult to coordinate, but industry respondents appeared to exhibit less interest in conducting distributed benchmarks than academic respondents.
- Industry respondents' processes tended to focus on short-length, single-client workloads, whilst academic respondents emphasised the importance of longer-running and mixed-client workloads.

2.3 Further Analysis of Benchmarking Processes

Our interviews and our own experiences suggested that benchmark execution can be represented as a *pipeline* of three main processes: (i) **initial configuration**, (ii) **benchmark runs**, and (iii) **results processing/analysis**. This pipeline model, depicted in Fig. 1, is consistent with observations of others in the community [5] who also approach benchmarking as a pipeline, albeit from a different perspective.

In Table 1 we assign each identified pain point to one or more pipeline processes. We observe that *initial configuration* and *results processing/analysis* are the two major sources of pain and time consumption, potentiating the greatest improvements in efficiency and usability for the overall benchmarking pipeline. Hence, our new benchmarking framework proposed in Sect. 3 is motivated by improving the efficiency of the pipeline by finding solutions to these pain points.

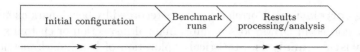

Fig. 1. Benchmarking represented as a pipeline. *Initial configuration* and *results processing/analysis* are the most painful processes, potentiating the largest efficiency and usability improvements (indicated by inward arrows).

Table 1. Association of identified pain points with benchmark pipeline processes.

Pipeline process	Major pain points
Initial configuration	PP1, PP2, PP3, PP4, PP5, PP6
Benchmark runs	PP4, PP5
Results processing/analysis	PP4, PP5, PP6, PP7, PP8

3   New Benchmarking Framework

In this Section, we propose and envision a new framework towards addressing the major pain points described in Sect. 2, with the overall goal of decreasing the inefficiency and pain associated with DBMS benchmarking. The framework we envision is founded on several key principles:

1. **Generality and versatility**—no restrictions should be imposed regarding the: benchmark; configuration parameters (DBMS/benchmark); target DBMS; workload; dataset; experimental method; or outputs. A relevant benchmarking framework should be capable of handling the heterogeneity of modern DBMS benchmarking.
2. **Extensibility and abstraction**—the framework should be highly extensible in all directions, with a modular design utilizing multiple abstraction layers. It should be straightforward to implement or extend benchmarks, workloads, target DBMSes, experimental methodologies, etc. This principle responds to the rapid pace of development within the database community and it aims to ensure the proposed framework's ongoing relevance.
3. **Usability and configurability**—the framework should be simple to install, configure, and run, painless to extend, and provide intuitive output. All aspects of the benchmarking pipeline should be separately and extensively configurable using a simple self-documenting configuration format. Running traditionally complex distributed benchmarks should be as simple as specifying a few configuration parameters.
4. **Track everything**—as much information as practically possible (i.e. without interfering with results) should be collected about the benchmarking environment. Metadata about relationships between information and objects within the system should also be collected. More context is better than less when reviewing benchmarking outputs.
5. **Repeatability/reproducibility**—replicating an experiment for result verification and consistency should be as simple as running a command.

6. **Flexibility of output**—the framework should be capable of outputting any well-defined format specified by the user, and of re-writing output in different formats after experiment completion. It should be straightforward to extend the framework with custom output formats.

Through consistent focus on these key principles, our proposed framework would make leaps in resolving the major pain points identified in Sect. 2.2. The remainder of Sect. 3 envisions our new framework through these key principles, and explains how each of the major pain points would be addressed. Though not the focus of this paper, we also developed a set of nonfunctional requirements and UML class diagrams for our proposed framework; these are available online[2].

3.1 Versatility and Extensibility

We propose a highly modular design suitable for implementation in any object-oriented programming language. In particular, our design supports any industry-standard benchmark or micro-benchmark. These could be implemented natively within our framework or run as separate programs. The minimum implementation required to run an existing benchmark would be writing methods to launch the existing benchmark's executable, process its output, and handle its input configuration parameters. The framework would similarly support any possible target DBMS, either implemented natively or accessed via a separately-running benchmark program.

Different experimental methods and repeatability (PP3) would be supported by abstracting the concept of a benchmark from that of an experiment. In our model, an experiment could use any benchmark or combination thereof, any number of times, with any number of warm-up/cool-down period, termination, and data collection triggers. This would allow expressing complex experiments such as the hypothetical scenario in Table 2.

Since our framework would be capable of handling the whole benchmarking pipeline, the need to write and debug custom testbed scripts (PP1–PP4) would be completely eliminated. Experiments designed within our framework could be easily ported to new scenarios without the traditional script-modifying and re-debugging that would otherwise be required with a custom testbed. Our framework would also provide new opportunities for collaboration and data sharing because anyone who could run the framework could also load and explore the output of any experiment performed using it.

3.2 Configuration

Simplifying configuration of benchmarks and DBMSes would be a significant usability accomplishment. Many systems are configured by setting values for a set of predefined configuration keys, often using a key-value configuration format such as Java `Properties` files. We would take advantage of this commonality

---

[2] https://github.com/lexibrent/benchfw-resources/.

Table 2. Hypothetical YCSB benchmarking experiment in our framework.

Benchmark config	`ycsb/workloads/workloada`
Benchmark clients	localhost
DBMS servers	bench1
Method	5 trials, non-distributed
Vary	YCSB thread count from 1 to 32, stepping by 1
Targets	MongoDB
Warm-up until	server disk I/O is stable
Output	YCSB throughput and aggregates: avg, SD, min, max
Output formats	`CSV` and `JSON`
Collect (from servers)	disk, CPU, RAM, and network utilization every 2 s
Start condition	start at `2018-02-10 00:00:00 UTC`

with a simple self-documenting key-value configuration format for every aspect of the framework. Our framework could automatically generate configuration files for other software components based on values set in the framework's own configuration, provided the framework is first extended with an implementation of the appropriate parse and generation logic (for non-key-value formats).

Any configuration file within our proposed framework can reference any other configuration file, allowing large or complicated scenarios to be split into manageable chunks. The need to copy entire configuration files to change options between experimental runs (PP1) is eliminated because our framework would allow all desired values to be expressed within the same configuration file using, for example, a concise range syntax. These value set declarations would be processed independently to the general configuration syntax, allowing custom syntax and parse logic to "just work" when implemented as an extension.

In addition, settings could be configured at different granularity levels. In order of granularity from course to fine, these would be: framework, DBMS, benchmark, experiment, and run. Any key configured at a given granularity level would override any values set for it in courser granularity levels. This design would allow, e.g., multiple experiments to share a common base configuration, with different sets of benchmark runs overriding specific configuration values.

Since flexibility is a major goal of our proposed framework, every component would be extensively configurable. Resource-intensive components such as comprehensive real-time system monitoring could be readily disabled. This design provides finer control of overhead and the associated trade-offs.

3.3 Distributed Benchmarks

Challenges associated with manually coordinating distributed benchmark clients running in parallel (PP5) would be eliminated by our framework's ability to automatically configure and coordinate multiple DBMS server systems and benchmark clients. An experiment configuration could list multiple machines on the same network that are all running the framework, partitioned into DBMS server and benchmark client machines. If any machine in such a configuration contains a more recent version of an experiment configuration than its peers, all peers would update to the later version before beginning execution. This would enable painless configuration modifications post-deployment, by simply modifying the configuration stored on one of the connected machines.

Partitioning the benchmark workload could be accomplished by specializing specific configuration values for each client or group of clients. For example, to split the workload based on operation type, one group of clients could be configured to perform only reads and another to perform only inserts and updates. Any configuration options supported by the benchmark could be used, so other examples may include partitioning based on operation count, primary key, table, or database (for multi-tenant benchmarks).

Our framework would track time stamps for all captured events, including error log entries and output from the benchmark and target DBMS. At the conclusion of a distributed benchmark run, all machines running the framework would collate these events based on times tamp, while still tracking which machine captured each event. Correlation coefficients between different datasets could be automatically computed for collected metrics, assisting the user with post-experiment analysis and debugging unexpected results.

3.4   Repeatability/Reproducibility and Debugging

Despite being a core principle of the scientific method, reproducibility is often overlooked by the database community. We speculate that this may be due to factors such as experimental complexity, inability to replicate hardware configurations, closed-source or proprietary software licences, and incomplete or imprecise descriptions of experiments in literature. Our proposed framework would simplify reproducibility (PP1, PP3) for benchmarking by automatically configuring the experimental method, benchmark, and DBMS. Each DBMS under test would simply need a corresponding adapter class implementation.

Using our framework, anyone with the necessary hardware and software environment could replicate any previous experiment performed with the framework by running a command and providing the output data from the experiment to be replicated. Our framework would warn the user if any detectable differences were found between the current environment and the environment used in the original experiment, reducing the chance of small differences going unnoticed. If the full output of a previous experiment is unavailable, the experiment could still be replicated by using identical configuration files, in which case our framework would be unable to report environmental differences. During replication of an experiment, values that were originally sampled from statistical distributions could either be re-sampled or re-used.

The comprehensive targeted metadata collection capabilities of our framework (described in Sect. 3.5) would simplify debugging by providing a more configurable level of detail than what is traditionally available. Debugging tools could also be attached to processes within the pipeline by setting appropriate configuration keys. This would eliminate the need to repeat an experiment with debugging tools manually attached, making heisenbugs easier to catch (PP4).

3.5   Metadata Collection

The framework's general approach to environment metadata collection would be that "more is better" provided it can be used effectively. So, it should be traceable, comprehensive, and relevant. Environment information would be collected for each system running our framework within an experiment. It would include details of (for example) the kernel, operating system, hardware, resource use, runtime and library versions, network connections, running processes and threads, system calls, memory access violations, crash dumps, and exact configuration files used. The collection mechanism would be modular and extensible, so additional data collection could be implemented with ease.

Traceability of collected metadata would be accomplished with a relational model linking each piece of information to its origin and other related data. For example, system environment information is related to the: machine on which it was collected, benchmark being executed, configuration used in that benchmark execution, experimental method used and its configuration, framework version, etc. These proposed metadata collection capabilities offer significant usability and efficiency improvements over typical benchmarking methodologies (PP6).

3.6   Output and Analysis

Industry-standard DBMS benchmarks output a variety of different formats, only some of which are extensible and configurable. Many write directly to stdout, relying on the user to manually perform collection and analysis (or automate it with custom testbed scripts). Our framework would improve this (PP2, PP4) by automatically extracting metrics of interest from raw benchmark output and storing these internally so they can be exported to any desired format such as CSV, JSON, or perhaps a plot. For each supported benchmark, our framework would require logic to process the benchmark's raw output; typically involving running regular expressions or parsers over captured stdout. Each output format would have its own writer implementation, so it would be straightforward to extend the framework with support for new output formats.

Comprehensive data collection in our proposed framework would make it possible to extract additional metrics from existing experimental output without needing to re-run the benchmark itself. This could also be used to convert a completed experiment's data to a different output format.

Statistical analysis of experimental results is traditionally performed entirely manually after data extraction (PP8). Our framework would improve the efficiency of this process by automating some common calculations. For example, the user could configure the framework to output aggregate metrics such as sum, average, min, max, median, standard deviation, variance, confidence interval, and correlation coefficients (PP7). As with the other aspects of our framework, these would be modular and extensible. The framework would also be capable of reporting new correlations that may be of interest to the experimenter.

4 Future Work

We now need to implement and evaluate our proposed framework. Future work could also explore developing a framework supporting both synthetic and trace-based workloads; combining the framework described in this paper with the functionality of BenchFoundry [3].

5 Conclusion

In this paper we proposed a new versatile and extensible framework for conducting benchmarking of DBMSes, based on a survey of the benchmarking practices of several individuals in the database community from both industry and academia. We showed that a typical benchmarking workflow is well-modelled as a pipeline of three key processes. Based on interview responses, we developed a set of major benchmarking "pain points" and mapped these onto the processes in our pipeline to determine which processes potentiated the greatest overall efficiency and usability improvements. We characterized several core principles upon which our vision is based: extensibility, usability, configurability, extensive data collection, and reproducibility. Our proposed framework was described, including how it would address each of the major pain points we had identified. A future implementation of our proposed framework could greatly improve the coherence of benchmarking for industry and academic purposes.

References

1. Ameri, P., Schlitter, N., Mayer, J., Streit, A.: NoWog: a workload generator for database performance benchmarking. In: 2016 IEEE 14th International Conference on Dependable, Autonomic and Secure Computing, 14th International Conference on Pervasive Intelligence and Computing, 2nd International Conference on Big Data Intelligence and Computing and Cyber Science and Technology Congress, DASC/PiCom/DataCom/CyberSciTech 2016, Auckland, New Zealand, 8–12 August 2016, pp. 666–673 (2016)
2. Barahmand, S., Ghandeharizadeh, S.: D-Zipfian: a decentralized implementation of Zipfian. In: Proceedings of the Sixth International Workshop on Testing Database Systems, DBTest 2013, pp. 6:1–6:6. ACM, New York (2013)
3. Bermbach, D., Kuhlenkamp, J., Dey, A., Ramachandran, A., Fekete, A., Tai, S.: BenchFoundry: a benchmarking framework for cloud storage services. In: Maximilien, M., Vallecillo, A., Wang, J., Oriol, M. (eds.) ICSOC 2017. LNCS, vol. 10601, pp. 314–330. Springer, Cham (2017). https://doi.org/10.1007/978-3-319-69035-3_22
4. Bermbach, D., Kuhlenkamp, J., Dey, A., Sakr, S., Nambiar, R.: Towards an extensible middleware for database benchmarking. In: Nambiar, R., Poess, M. (eds.) TPCTC 2014. LNCS, vol. 8904, pp. 82–96. Springer, Cham (2015). https://doi.org/10.1007/978-3-319-15350-6_6

5. Bermbach, D., Wittern, E., Tai, S.: Cloud Service Benchmarking. Springer, Cham (2017). https://doi.org/10.1007/978-3-319-55483-9
6. Cooper, B.F., Silberstein, A., Tam, E., Ramakrishnan, R., Sears, R.: Benchmarking cloud serving systems with YCSB. In: Proceedings of the 1st ACM Symposium on Cloud Computing, SoCC 2010, pp. 143–154. ACM, New York (2010)
7. Dey, A., Fekete, A., Nambiar, R., Rohm, U.: YCSB+T: benchmarking web-scale transactional databases. In: Proceedings - International Conference on Data Engineering, pp. 223–230 (2014)
8. Difallah, D., Pavlo, A.: OLTP-bench: an extensible testbed for benchmarking relational databases. Proc. VLDB Endow. **7**(4), 277–288 (2013)
9. Ghazal, A., et al.: BigBench: towards an industry standard benchmark for big data analytics. In: Proceedings of the ACM SIGMOD International Conference on Management of Data, SIGMOD 2013, New York, NY, USA, 22–27 June 2013, pp. 1197–1208 (2013). https://doi.acm.org/10.1145/2463676.2463712
10. Hoag, J.E., Thompson, C.W.: A parallel general-purpose synthetic data generator. SIGMOD Rec. **36**(1), 19–24 (2007)
11. Lu, J.: Towards benchmarking multi-model databases. In: 8th Biennial Conference on Innovative Data Systems Research, CIDR 2017, Chaminade, CA, USA, 8–11 January 2017, Online Proceedings (2017)
12. Rabl, T., Frank, M., Sergieh, H.M., Kosch, H.: A data generator for cloud-scale benchmarking. In: Nambiar, R., Poess, M. (eds.) TPCTC 2010. LNCS, vol. 6417, pp. 41–56. Springer, Heidelberg (2011). https://doi.org/10.1007/978-3-642-18206-8_4
13. Rabl, T., Poess, M., Danisch, M., Jacobsen, H.A.: Rapid development of data generators using meta generators in PDGF. In: Proceedings of the Sixth International Workshop on Testing Database Systems, DBTest 2013, pp. 5:1–5:6. ACM, New York (2013)
14. Sakr, S., Casati, F.: Liquid benchmarks: towards an online platform for collaborative assessment of computer science research results. In: Nambiar, R., Poess, M. (eds.) TPCTC 2010. LNCS, vol. 6417, pp. 10–24. Springer, Heidelberg (2011). https://doi.org/10.1007/978-3-642-18206-8_2
15. Seybold, D.: Towards a framework for orchestrated distributed database evaluation in the cloud. In: Proceedings of the 18th Doctoral Symposium of the 18th International Middleware Conference, Middleware 2017, pp. 13–14. ACM, New York (2017)
16. Stephens, J.M., Poess, M.: MUDD: a multi-dimensional data generator. SIGSOFT Softw. Eng. Notes **29**(1), 104–109 (2004)
17. Transaction Processing Performance Council (TPC): TPC-Homepage V5 (2016). http://www.tpc.org/
18. Van Aken, D., Difallah, D.E., Pavlo, A., Curino, C., Cudré-Mauroux, P.: BenchPress: dynamic workload control in the OLTP-bench testbed. In: Proceedings of the 2015 ACM SIGMOD International Conference on Management of Data, SIGMOD 2015, pp. 1069–1073. ACM, New York (2015)
19. Yoon, D.D.Y.: Database Performance Evaluation Framework. Ph.D. thesis, The University of Sydney (2008)
20. van der Zijden, W., Hiemstra, D., van Keulen, M.: MTCB: a multi-tenant customizable database benchmark. In: Proceedings of the 9th International Conference on Information Management and Engineering, ICIME 2017, pp. 17–23. ACM, New York (2017)

Neighbourhood Blocking
for Record Linkage

Daniel Elias[1]([✉]) and Josiah Poon[2]

[1] Commonwealth Bank, Sydney, Australia
daniel.elias@cba.com.au
[2] University of Sydney, Sydney, Australia
josiah.poon@sydney.edu.au

Abstract. This paper describes Neighbourhood Blocking – a novel method for the indexing step in the record linkage process. Record Linkage is the task of identifying database records referring to the same entity without the aid of definitive key fields. It has applications in data integration, fraud detection and other areas. This involves comparing pairs of records. If done indiscriminately, the size of this task is quadratic in dataset size. Therefore, various indexing methods are typically used to reduce the number of record pairs subjected to detailed comparison. Neighbourhood Blocking generalizes two existing indexing methods – Standard Blocking and Sorted Neighbourhood Indexing. It also allows meaningful treatment of missing values and a limited number of blocking field mismatches. Comparison of the Cartesian product of the blocks is avoided through the use of recursion. Numerical experiments and tests on benchmark datasets are reported in which Neighbourhood Blocking is compared to Standard Blocking and Sorted Neighbourhood Indexing. Under the conditions tested, Neighbourhood Blocking is found to frequently produce superior index quality, often at the expense of increased runtime. Scale testing indicates that index production speeds for Neighbourhood Blocking and Standard Blocking are similar when the database size is sufficiently large.

1 Introduction

1.1 Record Linkage

Many applications such as data integration, deduplication and fraud detection require the identification of distinct records referring to the same entity without the aid of unambiguous identifying fields. For example, many census datasets lack a field suitable for unambiguously identifying individual people. In such cases, the task of resolving object identity must rely on a combination of other field values where each field provides a partial and imperfect indication of object identity. In the case of census data, these fields might include name, address, date of birth, country of birth and other personal details. These non-key fields typically have a many-to-many relationship with object identity. To take address

© Springer Nature Switzerland AG 2019
L. Chang et al. (Eds.): ADC 2019, LNCS 11393, pp. 57–78, 2019.
https://doi.org/10.1007/978-3-030-12079-5_5

as an example, several people might live at the same address, and the same person may have different addresses at different times (or even a choice of homes at the same time). There can also be multiple representations of the same piece of information due to variations in the use of abbreviation, preferred names (e.g.: Betty rather than Elizabeth), spelling of numbers etc. Additionally, some differences are simply due to errors.

Since [3] investigated record linkage in 1940s, the subject has been pursued separately in several disciplines. Consequently, many names are now used to refer to it. These include: conflation, coreference resolution, data linkage, data matching, deduplication, disambiguation, entity resolution, name resolution, object identification, propensity score matching, record linkage, reference reconciliation and several others. In this paper, it will be referred to generally as "record linkage" and as "deduplication" when the records to be matched are in the same table.

The steps in the record linkage process described by [2] are:

1. *Preprocessing* – extraction of normalized representations and other features from the source dataset(s)
2. *Indexing* – Selection of record pairs for further consideration as possible matches
3. *Comparison* – Feature extraction from the selected record pairs
4. *Classification* – Binary classification of record pairs into matches and non-matches
5. *Evaluation* – Assessment of the quality of the record linkage produced

In deduplication, *merging* is often added as a final step. This involves producing a single record from each group of records that refer to the same entity.

This paper focuses on the *Indexing* step, the purpose of which is to inexpensively eliminate the vast majority of possible record pairs from consideration as possible matches. To illustrate why this is desirable, consider the deduplication of the dataset illustrated in Table 1. A rapid, high-recall indexing step with only 0.5% precision would reduce the number of record pairs to be considered in detail from approximately 5 billion to 1 million - a factor of $\frac{1}{5,000}$.

Table 1. Example dataset for deduplication

Item	Count
Total records	100,000
Duplicate records	5,000
Total record pairs	4,999,950,000
True matching record pairs	5,000

Table 2. Common indexing techniques

Indexing technique	Record pair selection criterion
Standard blocking	Exact match in one or more fields
Sorted neighbourhood	Proximity in a sorting order
Qgram-based, suffix-based	Exact match in any slight variations of a field's value
Canopy clustering; many: 1 mappings	cluster or group membership

The indexing step involves several competing objectives:

– Scalability - sufficient to accommodate the source dataset
– Size reduction - elimination of a large proportion of possible record pairs. This proportion is called the "reduction ratio".
– Recall - proportion of true matching pairs retained

Common indexing techniques are listed in Table 2. These are described more fully in Sect. 2.

1.2 Motivation

The primary motivation for combining aspects of Standard Blocking and Sorted Neighbourhood Indexing is to produce a single indexing method with the following features:

– Matching by "proximity" of field values (in addition to equality)
– Use of multiple fields to assess record similarity

As outlined in sections "An Issue with Standard Blocking" and "An Issue with Sorted Neighbourhood Indexing", Standard Blocking lacks the first of these features and Sorted Neighbourhood Indexing lacks the second.

An Issue with Standard Blocking. Standard Blocking is similar to a database join in that it matches record pairs where the values of certain specified fields ("blocking keys") are equal. A strength of this approach compared to Sorted Neighbourhood Indexing is that it simultaneously constrains the differences between values in multiple fields.

However, one disadvantage of Standard Blocking is that the only selection criterion for record pairs is whether or not they are in the same block. Any meaningful notions of block proximity or position *within* blocks (e.g.: when the blocks are discretized versions of continuous variables) are ignored. Take for example, the eight points illustrated in Fig. 1. Ideally, an indexing method that includes pairs of points that are far apart (for example, pair AD) should also include all pairs of points in the central cluster (i.e.: DE, EF and DF).

Depending on the positions of block boundaries, this will happen in some cases but not others. For example, the block boundaries in Fig. 2 separate all the points in the central cluster.

Fig. 1. Example points for indexing illustrations

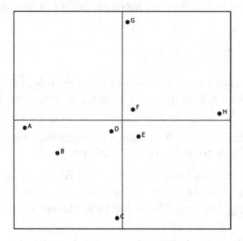

Fig. 2. Bivariate blocks

An Issue with Sorted Neighbourhood Indexing. Sorted Neighbourhood Indexing produces an ordering of the distinct combinations of values in certain fields ("sorting keys") and returns all record pairs whose combinations of sorting key values are closer than a specified distance in that ordering. Since the "windows" used to select records for pairing can overlap, the problem (described above) of nearby pairs straddling block boundaries doesn't apply in Sorted Neighbourhood Indexing.

However, one weakness of Sorted Neighbourhood Indexing is that it uses only a single ordering of the records, causing it to typically include more distant record pairs than, say, Standard Blocking with multiple blocking keys.

To illustrate this, consider once again the set of points in Fig. 1. Any ordering of the points corresponds to a (one-dimensional) route that visits all the points. For example, sorting by the horizontal ordinate and then the vertical one produces a route that is monotonically non-decreasing in the horizontal ordinate and varying widely in the vertical one. This is illustrated in Fig. 3. Taking a window size of 3 produces point groups ABD, BDC, DCG, CGF, GFE and FEH. The Sorted Neighbourhood Index (i.e.: all intra-group pairs of records) therefore includes distant pairs such as DC, GF and CG while excluding near ones such as DF and DE. CG is an example of a more distant pair than any that would be included by bivariate Standard Block Indexing as illustrated in Fig. 2.

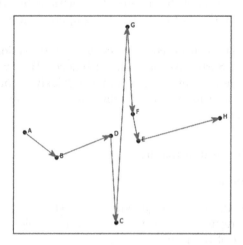

Fig. 3. Sorted Neighbourhood: route with sorting by horizontal variable

Other Issues

Treatment of Missing Values. Neither Standard Blocking nor Sorted Neighbourhood Indexing provide any meaningful treatment of missing values. When these methods are used in practice, missing values are typically either imputed or excluded (by omitting either rows or columns from the tables being linked). However, in reality missing values are just that – missing. The fact that a value is missing from a particular record neither confirms nor contradicts that record's pairing with any other record. Therefore, some allowance for limited wildcard matching of missing values is often desirable.

Allowance for Limited Field Mismatches. Records that refer to the same entity often contain differences in field values that would place them far apart in the sorting order. For example, consider the records shown in Table 3. These appear to refer to two individuals - Catherine and Timothy Bourke, each of whom is referred to by a pair of records. In each pair of records, there are three fields which

Table 3. Example records for deduplication

Row	Given name	Surname	Address	Birth date	SSID
1	Catherine	Bourke	42 Black Stump Cres	15-Mar-1958	3984257
2	Cathy	Smythe	42 Black Stump Cres	15-Mar-1958	398425
3	Timothy	Bourke	42 Black Stump Cres	06-Dec-1959	3939872
4	Timothy	Bourk	110 Beachfront Drive	06-Dec-1995	3939872

either agree exactly or would be nearby in the field's sorting order. However, they aren't the same three fields in both cases. In situations like this, a record pair selection rule like "any three of these five fields approximately agree" would be desirable.

Indexes of this type can be constructed using intersections and unions of Standard Blocking or Sorted Neighbourhood Indexes. However, the number of separate indexes to be computed and combined quickly becomes unwieldy. An algorithm that provides this type of matching more directly can be of value in situations like this one.

1.3   This Paper's Contributions

This paper's contributions are:

- Proposal and description of Neighbourhood Blocking
- Description of a recursive implementation which avoids comparison of the Cartesian product of the records
- Theorems relating to:
  - Inclusion of all pairs of records that are closer to one another than a specific Euclidean distance.
  - Conditions under which Neighbourhood Blocking is an unambiguous superset of Standard Blocking
  - Neighbourhood Blocking Index size relative to that of Standard Blocking
- Tests comparing Neighbourhood Blocking, Standard Blocking and Sorted Neighbourhood Indexing with respect to:
  - index quality in benchmark datasets
  - scalability properties in randomly generated datasets

1.4   Terms and Abbreviations

Key. A field used for grouping records into Blocks (blocking key) or for sorting them (sort key)

BKV. Blocking Key Value - one of the values contained in a blocking key

Block. A group of all records with a particular combination of BKVs

2   Related Work

2.1   Overview

The purpose of indexing in record linkage is to produce pairs of records for further consideration in the Comparison and Classification steps. The need for this was identified in [4] where the indexing technique now known as Standard Blocking is also described. Indexing for record linkage is a field of active research, and several approaches described in this section have significant similarities to Neighbourhood Blocking.

2.2   Full Index

The simplest way of selecting record pairs for further consideration is simply to select all possible pairs. This is known as "Full Indexing" and produces indexes of the sizes indicated in Table 4. The full index size can be manageable in the case of smaller datasets. For example, [8] focuses on the Comparison phase of the record linkage process (i.e.: comparisons of pairs of records rather than the selection of pairs for consideration). Numerical deduplication experiments are performed there on datasets with 2,000 rows and therefore approximately 2,000,000 record pairs in a full index.

Table 4. Full index sizes

Task	Full index pair count
Linkage (n rows to m rows)	nm
Deduplication (n rows)	$\frac{n^2 - n}{2}$

2.3   Standard Blocking

Standard Blocking, described by [4], produces all pairs of records which have exact matches in all fields designated as "blocking keys" (there can be multiple blocking keys). This can be implemented using the following steps:

1. Produce an inner join of the table(s) using the blocking keys (for deduplication, there is only one table - join it to itself). Retain only the two columns containing the record identifiers for the left and right tables in the join.
2. If there is only one source table (i.e.: the index is for deduplication rather than linkage), discard all rows where the left row identifier is greater than or equal to the right row identifier.
3. The remaining pairs of row identifiers are the index.

2.4 Sorted Neighbourhood Indexing

This method, described in [5] selects all pairs of records that are within a fixed "rank distance" of one another in a (single) sorted list. Unlike Standard Blocking, Sorted Neighbourhood Indexing produces the union of full indexes of *overlapping* groups of records. Since these record groups overlap, Sorted Neighbourhood Indexing avoids the block boundary problem described in Sect. 1.2. [9] describes an efficient method for implementing an online version of Sorted Neighbourhood Indexing using a tree-based approach to maintain the sort order as new records are added.

Sorted Neighbourhood Indexing requires two parameters:

- a record sorting criterion, and
- a "window width" which must be an odd positive integer

These steps can be used to produce a Sorted Neighbourhood Index:

1. Produce a single table of all distinct combinations of sorting keys in the dataset (i.e.: if there are two tables, take the union of the sorting key combinations in both of them). Call this the "Key Combination Table"
2. Sort the Key Combination Table.
3. For each record in the source table(s) find which row in the Key Combination Table has the same combination of sorting keys. Store these row numbers in a new "Rank" column in the source table(s)
4. Join the source table(s) (if there's only one then join it to itself). The join condition is that the absolute difference in Rank does not exceed $\frac{w-1}{2}$ where w is the window width. Retain only the two columns containing the record identifiers for the left and right tables in the join.
5. If there is only one source table (i.e.: the index is for deduplication rather than linkage), discard all rows where the left row identifier is greater than or equal to the right row identifier.
6. The remaining pairs of row identifiers are the index.

2.5 Mappings and Value Modifications

These methods map individual blocking key values to one or more alternative versions.

Many-to-one mappings (such as Soundex which maps strings to sound codes) are a way of coarsening blocking (thereby increasing the number of record pairs included).

Many-to-many mappings (such as string modifications using q-grams or suffix arrays) result in blocking where each record is effectively a member of multiple blocks, thereby reducing the effects of block boundaries. Some such methods are described in chapter four of [2]. Many-to-many mappings can often result in very large indexes. [1] describes an approach to address this by "pruning" the size of the mapping. This involves including only relatively rare BKV variants, greatly reducing the size of the index, but retaining those record pairs involving the coincidence of unusual variants.

2.6 Other Methods

Since indexing is applied to the entire source dataset, it typically has a greater need for scalability than the Comparison and Classification steps that follow it. Consequently, indexing algorithms typically require field comparison methods with properties that enhance algorithm scalability (e.g.: methods that rely on sorting require ordinal field comparisons).

Most indexing algorithms use these scalability-friendly comparison methods exclusively. However, some algorithms use them *in addition to* other comparison methods which lack the properties that support algorithm scalability.

Canopy Clustering is one such method. It is an approximate clustering method specifically suited to large datasets and distance functions that are slow to compute. It initially estimates distances between records using some faster distance function and then uses the slow distance function to revise distances below a threshold value. The steps involved in Canopy Clustering are:

1. Begin with no records allocated to clusters.
2. Choose an unallocated record at random. This will be the "centroid" of a new cluster. Use a fast comparison technique (e.g.: Jaccard similarity of q-gram sets) to identify other unallocated records that are similar to it (this involves comparison of the centroid record to all unallocated records). Then (only on those records selected) use *the slow comparison technique* to compute their distances to the centroid record. Allocate the centroid record and any others that were found to be sufficiently close to it to the new cluster.
3. As long as any records remain unallocated, continue repeating the previous step.

Another such method is Progressive Blocking which is described in [7]. This is a method of prioritizing the Comparison and Classification of record pairs with the aim of processing those more likely to be matches first. To do this, it relies on *integration with the Comparison and Classification tasks* in order to obtain feedback from them on the density of true matches found so far in different regions in the record pair space.

3 Neighbourhood Blocking

3.1 Intuition

Neighbourhood Blocking is a generalization of both Standard Blocking and Sorted Neighbourhood Indexing featuring:

- multiple blocking keys (like Standard Blocking)
- proximity (as well as equality) matching (like Sorted Neighbourhood Indexing)
- wildcard matching of missing values, and
- allowance for complete mismatches in a limited number of blocking keys

Since Neighbourhood Blocking allows proximity matching in *multiple blocking keys*, there is no single sorting order of the blocks such that block proximity corresponds to sorting order proximity. Therefore, some other method besides sorting needs to be used in order to avoid comparing the Cartesian product of the blocks. In the implementation described here, recursive application of the blocking algorithm achieves this.

3.2   Algorithm Description

Neighbourhood Blocking can be implemented in the following steps:

Normalize BKVs. Replace all non-null BKVs with integers representing their rank (i.e.: each non-null BKV is replaced with the count of *distinct* non-null values in the same column which appear earlier than it in a sorted list). This enables the coarsening and recursion described in step Sect. 3.2.

Atomic Blocking. Produce a (single) master table of block BKV combinations by taking all distinct combinations of BKVs in the table(s) being indexed. Assign a distinct block ID to each row in this table.

Produce a Linkage Index of Candidate Block Pairs. This is an index of the pairs of blocks on which the matching conditions will be checked in the next step. Neighbourhood Blocking is used (recursively) to achieve this as follows:

- If the blocking is maximally coarse (i.e.: each blocking key has only one non-null value), produce a Full Index (there will be no more than 2^n rows where n is the number of blocking keys).
- Otherwise: produce a Neighbourhood Blocking index using the same parameters (blocking keys, rank distance limits, wildcard limit and mismatch limit) on a coarsened version of the block table where each non-null BKV x is replaced with $\lfloor \frac{x}{a} \rfloor$ where $a > 1$.

The number of recursive steps is logarithmic in the maximum number of distinct values in any blocking key.

Identify Pairs of Matching Blocks. Compare the blocks in each of the block pairs identified in the previous step and determine which ones satisfy the record matching conditions. Put their block IDs into a link table (i.e.: each row is a pair of block IDs).

Translate Block Pairs to Record Pairs. Use database-style joins (via the link table found in the previous step) to determine the pairs of row IDs corresponding to matching row pairs.

For Deduplication, Filter the Record Pairs. In the case of deduplication of a single table (as opposed to linkage of two tables), filter the list of record id pairs to only include unique pairs (regardless of order).

3.3 Comparison to Other Methods

Standard Blocking and Sorted Neighbourbood Indexing are both special cases of Neighbourhood Blocking where no wildcard matching or match condition violations are allowed. Standard Blocking corresponds to a rank distance limit of zero (i.e.: only equality matching is allowed). Sorted Neighbourhood Indexing corresponds to a single (possibly composite) blocking key.

Strictly speaking, since Neighbourhood Blocking effectively involves a many-to-many mapping of BKVs, it bears some similarity to mapping-based techniques like q-grams or suffix arrays (described in Sect. 2.5). Neighbourhood Blocking differs from these other methods in that inexact BKV matches are restricted to those that can be determined from sorting order or null values alone.

Unlike Progressive Blocking (outlined in [7]), Neighbourhood Blocking is separable from the Comparison and Classification steps, uses multiple sorting orders to determine block proximity and allows for wildcard matching and field mismatches.

A comparison of some key features of Neighbourhood Blocking and some of its counterparts is summarized in Table 5.

Table 5. Comparison of index algorithm features

Feature	Standard	Sorted N'hood	Progressive	Neighbourhood
Multiple block keys	✓		✓	✓
Multiple orderings	N/A			✓
Block combination/overlap		✓	✓	✓
Separable from comparison	✓	✓		✓
Nulls as wildcards				✓
Limited non-matches				✓

3.4 Properties

The Proximity matching criterion allows the inclusion of record pairs which straddle block boundaries. By Theorem 1, where there is a notion of position within blocks (making the notion of "close pairs straddling block boundaries" meaningful), Neighbourhood Blocking includes all record pairs closer than a specific "inclusion distance", regardless of the specific locations of block boundaries.

Theorem 1 (Inclusion distance). *A Neighbourhood Blocking index using blocking keys that are discretized versions of continuous variables will include all pairs of records whose non-discretized Euclidean distance is less than the* *product of (a) the length of the unit of discretization, and (b) the lowest of its rank distance limits. This is true regardless of the locations of block boundaries.*

Proof. Let the differences in the values of each of the non-discretized blocking keys be $\delta_1, \delta_2 \cdots \delta_n$ (where n is the number of blocking keys). Let the unit of discretization be u and let the minimum of the rank distance limits be r_{min}. By assumption, the Euclidean distance between the records is less than $r_{min}u$. In other words:

$$\sqrt{\sum_j \delta_j^2} < r_{min}u \Rightarrow \sum_j \delta_j^2 < (r_{min}u)^2 \tag{1}$$

If the record pair is *not* included in the index by the Proximity criterion, the following condition must be true.

$$\exists j : \delta_j > r_{min}u \tag{2}$$

which implies:

$$\sum_j \delta_j^2 > (r_{min}u)^2 \tag{3}$$

since

$$\forall j : \delta_j^2 > 0 \tag{4}$$

Since (1) is a contradiction of (3), any pair of records separated by a Euclidean distance less than $r_{min}u$ must be in the index.

Clearly, a Neighbourhood Blocking Index is a superset of a Standard Blocking index which uses the same keys. However, by Theorem 2, it is also a superset of a Standard Blocking Index where the granularity of any or all of the blocking keys is coarsened by combining groups of $1 + r_j$ adjacent values where r_j is the jth blocking key's rank distance limit.

Theorem 2 (Superset of Standard Blocking). *If:*

1. *X_N is a Neighbourhood Blocking index with keys $k_1, k_2 \cdots k_n$ and corresponding rank distance limits of $r_1, r_2 \cdots r_n$*
2. *Each k'_j ($j \in \{1 \cdots n\}$) is a $(1 + r_j){:}1$ mapping of k_j such that each distinct value of k'_j corresponds to $(1 + r_j)$ consecutive sorted values of k_j*
3. *X_S is a Standard Blocking Index with keys $k'_1, k'_2 \cdots k'_n$*

Then: $X_N \supseteq X_S$

Proof. X_S comprises the union of Full Indexes on each of its blocks. Therefore, it suffices to show that the Full Index of each such block is included in X_N.

Consider any X_S block. All records in it contain identical values of $k'_1, k'_2 \cdots k'_n$. In the BKV matching for X_N, assumption 2 implies that the values for each of $k_1, k_2 \cdots k_n$ must match by the Proximity criterion. Thus, all record pairs in the X_S block are included in X_N.

Theorem 3 relates to an idealized database where:

- all keys have sufficiently many distinct values for edge effects to be negligible,
- records are uniformly distributed throughout the key space, and
- no keys have any null values

Under these idealized conditions, it is shown that the size of a Neighbourhood Blocking index is larger than a Standard Blocking Index using the same keys by a factor of $\prod_j(1 + 2r_j)$, where r_j is the rank distance limit for the jth blocking key. By Corollary 1, for the same idealized database and where all blocking keys are also sorting keys, a Neighbourhood Blocking Index has the same reduction ratio as a Standard Blocking Index with each key coarsened by a factor of $1+2r_j$.

Theorem 3 (Index size relative to Standard Blocking). *In datasets where each block contains the same number of records and each sorting key has the same number of distinct values, the index sizes for Standard Blocking and Neighbourhood Blocking are related by:*

$$\lim_{d,v \to \infty} \frac{|X_N|}{|X_S|} = \prod_j (1 + 2r_j) \tag{5}$$

where:

X_S *is the set of record pairs from Standard Blocking*
X_N *is the set of record pairs from Neighbourhood Blocking where no field mismatches or wildcards are allowed.*
r_j *is the rank distance limit used in Neighbourhood Blocking for the jth blocking key*
d *is the number of records per block,*
v *is the number of distinct values of each blocking key*

Proof. The ratio of the index sizes can be itemized into contributions from interior and non-interior blocks. Here, "interior block" means a block whose neighbourhood is not limited by maximum or minimum values of any of the sorting keys. The itemization of the index size ratio is expressed in (6).

$$\frac{|X_N|}{|X_S|} = p_I R_I + (1 - p_I) R_N \tag{6}$$

where:

p_I proportion of blocks that are interior blocks
R_I ratio of number of pairs contributed by interior blocks
R_N ratio of number of pairs contributed by non-interior blocks

The number of record pairs contributed to the Standard Blocking Index by every block (interior or not) is given by (7) for a deduplication index and by (8) for a linkage index.

$$\frac{d(d - 1)}{2} \tag{7}$$

$$d^2 \tag{8}$$

In Neighbourhood Blocking, each record is paired with the records in its own block and also any other block whose BKV ranks differ by no more than r_j for each j. A record in an interior block can therefore be paired with records in a total of $\prod_j (1 + 2r_j)$ blocks (since the paired record's ranking in the jth key's sort order can differ by any of: $-r_j \cdots 0 \cdots r_j$). Therefore, the number of record pairs contributed to the Neighbourhood Blocking Index by each *interior* block is given by (9) for a deduplication index and by (10) for a linkage index.

$$\frac{d\left(\left(\prod_j (1 + 2r_j)\right) d - 1\right)}{2} \tag{9}$$

$$d^2 \prod_j (1 + 2r_j) \tag{10}$$

R_I is given by (9) divided by (7) in the case of a deduplication index and by (10) divided by (8) in the case of a linkage index. In both these cases:

$$\lim_{d \to \infty} R_I = \prod_j (1 + 2r_j) \tag{11}$$

Let n be the number of blocking keys. The total number of blocks is v^n, and the number of interior blocks is $\prod_j (v - 2r_j)$ which is no smaller than $(v - 2r_{max})^n$ where $r_{max} = \sup_j r_j$ Therefore, the proportion of blocks that are interior blocks satisfies:

$$p_I \geq \left(1 - \frac{2r_{max}}{v}\right)^n \tag{12}$$

implying that:

$$\lim_{v \to \infty} p_I = 1 \tag{13}$$

Combining (13) and (11) with the recognition that R_N must be finite completes the proof.

Corollary 1. *In the limiting case described in Theorem 3, Neighbourhood Blocking produces the same index size as Standard Blocking with each blocking key coarsened by a factor of $(1 + 2r_j)$ where r_j is the jth blocking key's rank distance limit.*

Proof. Let the total number of records in the dataset be N. This is related to d, v and n by (14).

$$N = dv^n \Rightarrow d = Nv^{-n} \tag{14}$$

The size of a Standard Blocking index is therefore given by (15) for a deduplication index and by (16) for a Linkage index.

$$\frac{N(Nv^{-n} - 1)}{2} \tag{15}$$

$$N^2 v^{-n} \qquad (16)$$

If the blocks are coalesced into larger ones with a "side length" of $(1 + 2r_j)$ old blocks in the direction of each (j^{th}) blocking key, v^n in (14), (15) and (16) is effectively replaced with a smaller number $\frac{v^n}{\prod_j (1+2r_j)}$. Therefore, for both deduplication and linkage indexes in the limiting case as $N \to \infty$ the ratio of the size of a Standard Blocking Index with the larger blocks to that of one with the smaller blocks tends to:

$$\prod_j (1 + 2r_j) \qquad (17)$$

which is the same ratio given in (5).

3.5 Application to Earlier Examples

Figure 4 illustrates how the application of Neighbourhood Blocking addresses the two issues outlined in Sect. 1.2. Namely, application of proximity matching in multiple directions and inclusion of close pairs of points that straddle block boundaries. The rank distance limit is 1 for both keys, and the block size is one third that used in Fig. 2 (the proportion indicated by Corollary 1 for Neighbourhood Indexing to produce a similar index size to Standard Blocking).

The shading in Fig. 4 indicates blocks containing points that are paired with point D, the darker shaded block matching by Equality, and the lighter shaded ones matching by Proximity. Although this blocking contains boundaries in the same positions as those in Fig. 2 (and therefore the central cluster is still divided by block boundaries), these do not prevent the inclusion of point pairs from the central cluster in the index since all these pairs match by the Proximity criterion.

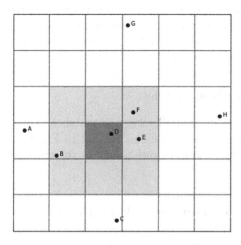

Fig. 4. Neighbourhood Blocking: pairings for point D

3.6   Wildcard Matching of Missing Values

Figure 5 illustrates the treatment of missing values in Neighbourhood Blocking. This includes the same points as in earlier examples, with the addition of two new ones:

- P which has a null value for the vertical (y) ordinate, and
- Q which has null values for both ordinates (x and y)

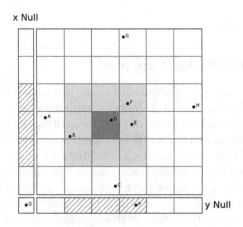

Fig. 5. Neighbourhood Blocking: pairings for point D allowing one missing value

The area highlighting in Fig. 5 indicates the matching criteria for point D when up to one (null value) wildcard match is allowed. Wildcard matches are indicated by the hatched areas in the bars containing null values. Since up to one wildcard match is allowed, pair PD will be included in the index but point Q (which has two null values) will not.

3.7   Allowance for Non-matches

Figure 6 illustrates the effect of allowing non-matches in one of the fields. In this figure, the blocking is twice as granular as that in Fig. 4. Allowing a mismatch in either of the two fields would cause points in the cross-hatched regions to be paired with point D.

4   Application to Benchmark Datasets

Comparisons of index quality (defined in Sect. 4.1) between Neighbourhood Blocking, Standard Blocking and Sorted Neighbourhood Indexing were made on the benchmark datasets listed in Table 6.

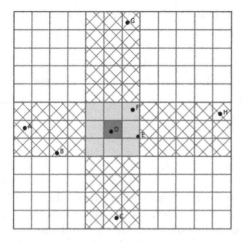

Fig. 6. Neighbourhood Blocking: pairings for point D allowing one non-match

Table 6. Sample datasets

Source	Dataset	Entity type	Column count(s)	Row count(s)	True matches	Calculated fields
Recordlinkage	FEBRL1	Person	13	1,000	500	Date components
	FEBRL2	Person	13	5,000	1,934	Date components
	FEBRL3	Person	13	5,000	6,538	Date components
	FEBRL4	Person	13; 13	5,000; 5,000	5,000	Date components
DBG Leipzig	Amazon-Google Products	Product	11; 11	1,363; 3,226	1,300	Parsed codes; topic modelling
	ABT-Buy	Product	10; 11	1,081; 1,092	1,097	Parsed codes; topic modelling
	DBLP-ACM	Publication	4; 4	2,616; 2,294	2,224	
	DBLP-Scholar	Publication	4; 4	2,616; 64,263	5,347	

4.1 Methodology

After the calculated fields described in Table 6 were added to the datasets, a number of indexes were calculated for each dataset using each of the indexing methods. These were based on valid combinations of the parameters listed in Table 7. For each index produced, a point representing its recall and reduction ratio was computed. These were grouped by indexing method and the "frontier points" among them were identified as those that:

1. are on the convex hull surrounding all points for the indexing method, and
2. do not have *both* lower recall and lower reduction ratio than any other point

Table 7. Indexing parameters for benchmark datasets

Variable	Values	
	Bounds	Distribution
Number of blocking keys	1 .. number of columns	Linear
Half window	1 .. $\frac{1}{4}$ number of rows	Geometric
Non-match limit	0, 1	

4.2   Results

Index quality frontiers for the tests performed are shown in Figs. 7, 8, 9, 10 and 11. Each figure relates to a dataset. Each curve within these figures relates to an indexing method. The points on the curve represent the frontier of combinations of recall and reduction ratio achieved with the corresponding indexing method and dataset.

The results for all the FEBRL datasets are broadly similar, so only Febrl3 is shown (Fig. 7). For all these datasets, index quality is moderate to high and the ranking of index quality frontiers is the same. That ranking is:

1. Neighbourhood Blocking with a near-perfect result
2. Sorted Neighbourhood Indexing
3. Standard Blocking

In the DBG Leipzig datasets (Figs. 8, 9, 10 and 11), all three indexing methods produce far lower index quality than those achieved in the FEBRL datasets. [6] reports using string similarity measures in the indexing step to achieve higher index quality in these datasets.

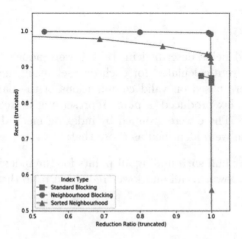

Fig. 7. Index quality frontiers - FEBRL3

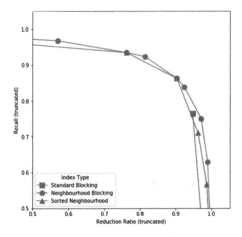

Fig. 8. Index quality frontiers - Abt-Buy

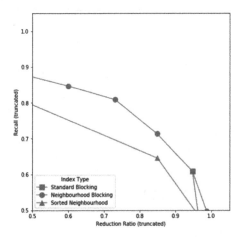

Fig. 9. Index quality frontiers - Amazon-GoogleProducts

In the case of the Amazon-GoogleProducts dataset (Fig. 9), Neighbourhood Blocking did produce a noticeable improvement over the other methods, but as with all the DBG Leipzing datasets, the absolute quality of all the indexes suggests either the use of a Full Index or a more general value comparison method (as used by [6]).

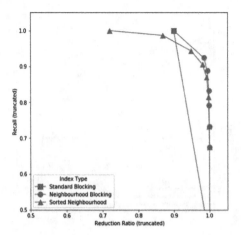

Fig. 10. Index quality frontiers - DBLP-ACM

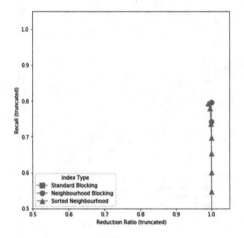

Fig. 11. Index quality frontiers - DBLP-Scholar

5   Scalability Comparison

Numerical experiments were conducted to compare the scalability of Neighbourhood Blocking with that of Standard Blocking and Sorted Neighbourhood Indexing on simulated datasets with up to 1 million rows. It was found that runtime
for Neighbourhood Blocking depends most strongly on two factors:

- How sparsely populated the blocks are (i.e.: the ratio of the number of records
  in the database to the number of distinct combinations of BKVs)
- The size of the final index produced

Table 8. Indexing times - lines of best fit by method (seconds per million row pairs)

Method	Filtering	Intercept	Slope	R^2
Neighbourhood Blocking - No wildcards or adjacency		4.92	3.18	0.33
Neighbourhood Blocking - no wildcards		8.68	3.12	0.08
Neighbourhood Blocking - 1 wildcard		9.23	3.02	0.09
Neighbourhood Blocking - 2 wildcards		10.27	3.15	0.07
Full		1.19	1.08	0.99
Sorted Neighbourhood		1.22	10.34	0.96
Standard Blocking		1.55	3.23	0.82
Neighbourhood Blocking	Non-sparse	0.26	3.20	0.99
Standard Blocking	Non-sparse	0.23	3.13	0.99

When database sparsity is low:

- runtime for all three indexing methods is approximately linear in the size of the index produced, and
- the rates of index production for Standard and Neighbourhood blocking are similar

5.1 Results

Table 8 shows relationships between index production times (in seconds) and index size (in millions of record pairs) for the methods tested. For non-sparse datasets (i.e.: those with many records per block), index production rates for Standard Blocking and Neighbourhood Blocking are similar.

6 Discussion

Since Standard Blocking and Sorted Neighbourhood Indexing are both special cases of Neighbourhood Blocking:

- Neighbourhood Blocking can produce the same indexes as the other two index types (as well as other indexes which might have higher quality), and
- The other two index types can be produced using the same algorithms as are used for Neighbourhood Blocking (as well as other, more restricted algorithms which might have lower resource consumption)

Therefore, whether or not Neighbourhood Blocking is preferable to the other two methods depends on whether it produces an increment in index quality that justifies any increment in resource consumption.

The benchmark datasets examined in Sect. 4 include several cases where the improvement in index quality is material (as well as some where all three methods behave poorly).

The timings in Sect. 5 indicate that the difference in resource consumption is larger for small, sparse datasets than for large, dense ones.

7   Conclusion

Compared to Standard Blocking and Sorted Neighbourhood Indexing, Neighbourhood Blocking can always produce indexes of at least the same quality, but will always require at least the same resources.

Compared to the other two methods, Neighbourhood Blocking has several advantages which can result in higher index quality. These include:

- multi-key proximity matching
- meaningful treatment of missing values
- tolerance for complete mismatches in a limited number of keys

Simple sorting-based implementations are inapplicable to the case of simultaneous proximity matching on multiple fields. However, efficient implementation is possible through use of recursion.

Scalability tests indicate similar index production speeds for Neighbourhood Blocking and Standard Blocking in sufficiently large datasets.

8   Further Work

This work could be extended by making a progressive version of Neighbourhood Blocking. This would be similar to Progressive Blocking as described by [7], except that additional match types would be allowed and block proximity would be determined by multiple sorting orders.

References

1. Khairul Nizam, B., Shahrul Azman, M.N.: Efficient identity matching using static pruning q-gram indexing approach. Decis. Support Syst. **73**, 97–108 (2015)
2. Christen, P.: Data Matching. Springer, Heidelberg (2012). https://doi.org/10.1007/978-3-642-31164-2
3. Dunn, H.L.: Record linkage. Am. J. Public Health Nation's Health **39**, 1412–1416 (1946)
4. Fellegi, I.P., Sunter, A.B.: A theory for record linkage. J. Am. Stat. Assoc. **64**, 1183–1210 (1969)
5. Hernandez, M.A., Stolfo, S.J.: The merge/purge problem for large databases. In: Proceedings of the 1995 ACM SIGMOD International Conference on Management of Data, San Jose, California, USA. ACM (1995)
6. Kopcke, H., Thor, A., Rahm, E.: Learning-based approaches for matching web data entities. IEEE Internet Comput. **14**(4), 23–31 (2010). https://doi.org/10.1109/MIC.2010.58
7. Papenbrock, T., Heise, A., Naumann, F.: Progressive duplicate detection. IEEE Trans. Knowl. Data Eng. **27**(5), 1316–1329 (2015). https://doi.org/10.1109/TKDE.2014.2359666
8. Poon, S.K., et al.: An ensemble approach for record matching in data linkage. Stud. Health Technol. Inform. **227**, 113–119 (2016)
9. Ramadan, B., Peter Christen, H.L., Gayler, R.W.: Dynamic sorted neighborhood indexing for real-time entity resolution. J. Data Inf. Qual. **6**(4), 15 (2015)

Items2Data: Generating Synthetic Boolean Datasets from Itemsets

Ian Shane Wong$^{(\boxtimes)}$, Gillian Dobbie, and Yun Sing Koh

Department of Computer Science, University of Auckland,
38 Princes Street, Auckland, New Zealand
iwon015@aucklanduni.ac.nz, g.dobbie@auckland.ac.nz, ykoh@cs.auckland.ac.nz

Abstract. Boolean data is a core data type in machine learning. It is used to represent categorical and transactional data. Unlike real valued data, it is notoriously difficult to efficiently design boolean datasets that satisfy particular constraints. Inverse Frequent Itemset Mining (IFM) is the problem of constructing a boolean dataset, satisfying given support constraints for some itemsets. Previous work mainly focuses on the theoretical complexity of IFM and practical solutions scale poorly or do not satisfy all the constraints. We propose Items2Data, a practical algorithm for generating boolean datasets which is efficient under specific conditions. We introduce global closure to describe the condition which a dataset can be efficiently constructed. We evaluate Items2Data and its use in designing synthetic datasets and to analyze its accuracy, scalability and speed on real world datasets. The results indicate Items2Data is practical and efficient for generating synthetic boolean data when predefined itemsets are globally closed.

1 Introduction

The generation of boolean data sets that satisfy particular constraints is needed to critically evaluate machine learning algorithms for categorical and transaction data. Popular matrix factorization approaches exist for designing and generating synthetic real valued datasets, such as the Cholesky Decomposition [1], but generating boolean data is a more difficult problem [2].

Various solutions have been proposed to solve the Inverse Frequent Itemset Mining (IFM) problem [3–7]. The ability to construct a dataset D from a set of itemsets S is powerful due to the intuitive nature of specifying itemset supports or frequencies. However, the IFM problem is intractable [8]. Previous work has explored many theoretical aspects of the problem and there are open questions as to whether there exists an efficient and practical solution. Of the practical solutions that exist [3–5] there is a trade-off between the ability to accurately reconstruct the support of itemsets in S and the time complexity of the algorithms.

Example: Consider an example where John wants to generate synthetic data for 2 different scenarios: (1) a, b, c are statistically independent of each other

© Springer Nature Switzerland AG 2019
L. Chang et al. (Eds.): ADC 2019, LNCS 11393, pp. 79–90, 2019.
https://doi.org/10.1007/978-3-030-12079-5_6

except for a slight positive association between a and b. (2) same as in (1), but additionally a and b are negatively associated in the presence of c (a complex interaction). He works out what that might look like in terms of itemset supports and comes up with the itemset properties in Table 1.

Table 1. Itemset properties for running example

Itemset	Independent support	Scenario 1 support	Scenario 2 support
a	0.500	0.500	0.500
b	0.500	0.500	0.500
c	0.500	0.500	0.500
ab	0.250	**0.300**	**0.300**
ac	0.250	0.250	0.250
bc	0.250	0.250	0.250
abc	0.125	**0.150**	**0.100**

Motivated by the ease of defining the joint distributions between items in a similar manner to defining the desired correlations between variables in a correlation matrix, we find a special condition that allows for the efficient reconstruction of a dataset. Under this condition we can efficiently reconstruct a dataset that satisfies itemset constraints in polynomial time.

We introduce concepts of marginal support, global closure and a novel approach to generating boolean datasets. The marginal support of an itemset is the non derivable information based on the support of all of its supersets. Calculating the marginal support for each itemset is similar to calculating how much of an itemset's support is covered by its supersets. By determining the marginal support we are able to detect all itemsets whose supports can be derived by their supersets. We find that the marginal support is related to the frequency of an itemset that will be present in a dataset. When the sum of all marginal supports in a set of itemsets is less than or equal to 1 then we say the set of itemsets is globally closed.

Closed itemsets [9] are a compressed representation of itemsets and are related to globally closed itemsets. The key difference between closed itemsets and globally closed itemsets is that closed itemsets are defined by whether any superset has the same support as an itemset and global closure is defined by whether the total marginal support of all supersets are the same as an itemset's support. With marginal support, multiple supersets in combination can cover an itemset not just a single superset. Globally closed itemsets are also a compressed itemset representation and are smaller or at least as small as closed itemsets.

Example: Consider the itemsets in Fig. 1 and their support. All itemsets are closed by definition. Itemset c is closed because it does not have the same support as any of its supersets, however c is derivable because its marginal support

Itemset	Support	Marginal	Marginal Formula
a	0.500	0.125	s(a) - m(ab) - m(ac) - m(abc)
b	0.500	0.125	s(b) - m(ab) - m(bc) - m(abc)
c	0.375	0.000	s(c) - m(ac) - m(bc) - m(abc)
ab	0.250	0.125	s(ab) - m(abc)
ac	0.250	0.125	s(ac) - m(abc)
bc	0.250	0.125	s(bc) - m(abc)
abc	0.125	0.125	s(abc)
{}	1.000	0.250	s({}) - m(abc) - m(bc) - m(ac) - m(ab) - m(c) - m(b) - m(a)

Fig. 1. A table of itemset support and marginal support. The marginal support is calculated using the marginal formula where s(I) is the support for each itemset and m(I) is the marginal support for each itemset.

is 0. This means the support of c can be derived from the marginal support of all its supersets. The minimal set of globally closed itemsets therefore excludes c and is smaller than the set of closed itemsets.

Using the properties of marginal itemsets and global closure, we propose an algorithm, Items2Data, which aims to reconstruct a dataset D that satisfies itemsets S. Taking a set of itemsets S with supports, we construct a set of marginal supports M. If M is globally closed we can derive a dataset D that satisfies the supports of M and therefore the supports of S. By focusing on the property of global closure, it has the potential to make reconstructing data from itemsets tractable. An advantage of this approach is that if S is not globally closed, then instead of attempting to solve the problem of reconstructing a dataset from S, we can change the problem to repairing S to make it globally closed. If S can be repaired efficiently then we open up the opportunity to efficiently generate boolean data much more generally.

Based on the above, our contributions in this paper are:

- The introduction of the concepts of marginal support and global closure, which are used to determine whether a set of itemsets and supports can efficiently map to a dataset.
- An efficient algorithm Items2Data that generates datasets that satisfy itemset constraints by taking advantage of marginal support and global closure.
- Novel application of IFM to design synthetic boolean data.

We run two sets of experiments. The first explores the construction of 2 synthetic environments introduced in Table 1 and use them to validate the data distributions generated by Items2Data under different classification algorithms. The second validates the reconstruction accuracy of Items2Data and time taken for reconstruction on 6 real world datasets. We also test the time taken to calculate the marginal support from itemset supports in each of the datasets, to show the practical feasibility of Items2Data.

The structure of the paper follows. Related work, alternatives to synthetic data generation and well understood concepts are described in Sect. 2. Section 3 introduces the novel concepts of marginal support, global closure and the Items2Data algorithm. We demonstrate and validate the effectiveness of Items2Data in Sect. 4, and conclude and describe future work in Sect. 5.

2   Related Work

In this section we review previous work in Inverse Frequent Pattern Mining (IFM), synthetic data generation and introduce some known concepts that provide the foundation for our work. The IFM problem was initially introduced in [10] and was analyzed from a theoretical perspective, showing that a general solution is NP-hard. This work was motivated by privacy concerns when itemset supports are shared publicly. Since then alternative formulations and parameterizations have been analyzed in [3,4,7,8], however it is not known whether there is an efficient method to solve IFM and its variants.

Motivated by the intractability of the IFM problem, several heuristics have been proposed. In particular [5] proposes an approximate solution to generating synthetic market basket data, using an Iterative Proportional Fitting method based on contingency tables. Guzzo et al. [4] introduced an alternative approach that relaxed support constraints from a fixed constraint into a minimum and maximum support. It also introduced the idea of satisfying the minimum support for as long as possible while guaranteeing the maximum support, which improves the tractability of the IFM problem. However, it is not guaranteed to satisfy itemsets S even if the minimum support is equal to the maximum.

In subsequent work Guzzo et al. [3] propose an alternative formulation where itemsets that are not in S are constrained to be infrequent below a threshold and solved using large scale linear programs. Ramesh et al. [6] generate a dataset that captures the properties of maximal itemsets, and does not aim to satisfy the supports of S exactly. Our focus is on practical solutions to the IFM problem. The key differences between previous practical solutions and our work is that we satisfy S exactly and in polynomial time as long as S is globally closed. We also ignore minimum support since it improves the likelihood of global closure.

Cholesky decomposition [1] is widely used to generate correlated multivariate random normal synthetic datasets. It is efficient to compute and design datasets by defining a valid correlation matrix, and the method works well for Gaussian distributions. However, the output of this method is a real valued matrix and it is not possible to transfer the method to boolean data generation. Matrix decomposition methods traditionally capture the pairwise correlation between variables and if interaction effects or conditional dependencies between variables are desired then it becomes non trivial to design a valid correlation matrix.

Generating discrete dependent random variables via Gaussian copulas [11] is another statistical alternative. Gaussian copulas can be used to define joint distributions for discrete variables. We propose an alternative approach using the itemset framework. By specifying the itemsets and supports a dataset must satisfy, we implicitly specify the joint distributions we wish to see. It is also simple to specify higher order relationships by defining an itemset containing several items.

2.1   Itemsets and Support

Itemsets are a well understood and fundamental concept in data mining, used to represent transactions [12].

Definition 1. *Let $\mathcal{I}$ be a finite domain of elements also known as items. Let $\{I_1, I_2, ..., I_n\}$ each be a unique collection of elements in $\mathcal{I}$ also known as itemsets. Let S be a set of itemsets with associated supports. A transaction contains an itemset $T = I_i$ and a dataset is a collection of transactions $D = \{T_1, T_2, ..., T_j\}$.*

TID	a	b	c
1	1	1	1
2	0	1	1
3	1	0	1
4	1	1	0
5	0	1	0
6	1	0	0
7	0	0	0
8	0	0	0

Fig. 2. A dataset with items a, b and c in transactions 1–8.

Consider the dataset that is described in Fig. 2, which has items a, b and c. Each row represents a single transaction and contains a single itemset. In transaction 1 items a, b, and c are all present so transaction 1 contains the itemset abc, while transaction 2 contains the itemset bc. The support of I_i is determined by $\frac{1}{|D|} \times |I_i \supseteq T|$ for $T \in D$, which is the fraction of transactions in a dataset that contain the itemset I_i. For example the support of a is 0.5.

2.2   Itemset Representations

We describe closed itemsets, maximal itemsets and non-derivable itemsets which are three different compressed itemset representations. Closed itemsets [9] have been popular as the set of all closed itemsets is a compressed itemset representation that can be used to reconstruct the support of every other itemset. An itemset is closed if none of its supersets have the same support as itself. Maximal itemsets [13] are a smaller itemset representation and can be used to reconstruct all itemsets, but not their supports. A non-derivable itemsets (NDI) representation [14] is a set of itemsets which are non-derivable. A derivable itemset is an itemset whose support can be perfectly derived by other itemsets, and all other itemsets are non derivable. That is all other sets and their supports can be reconstructed from an NDI representation.

3   Items2Data

We start by introducing new concepts before describing our novel Items2Data algorithm.

3.1   Marginal Support Representations

We will introduce what marginal support is and also present a simple algorithm to construct marginal support itemset representations from itemsets and their support. The marginal support is the additional support of an itemset after subtracting the marginal support of all its supersets. The intuition is that the support of an itemset is not only determined by the number of transactions that exactly match an itemset, but also by the number of transactions that are supersets of an itemset. In Fig. 2 for itemset a there is only 1 transaction that exactly matches a, but 3 other transactions with supersets of a giving a a support of 0.5. By subtracting the support that is covered by each superset, we can isolate the marginal support of a, that is the support information not derivable from all of its supersets ab, ac, abc.

The algorithm for computing the marginal support of a set of itemsets and their support can be solved in polynomial time. We wish to construct M where each element contains the marginal count of an itemset in S. First we must sort S by the length of the itemsets from longest to shortest resulting in S'. Starting from S'_1, the marginal support of S'_1 is the support of S'_1 minus the total marginal support of all its supersets. We repeat this process for $S'_2, ..., S'_n$. In the worst case, Quicksort is $O(n^2)$. There are also $n(n-1)$ superset checks to calculate all marginal counts which is also $O(n^2)$. The worst case complexity of computing marginals from S is $O(n^2)$ where $n = |S|$.

For example in Fig. 1, given $S = \{a : 0.500, b : 0.500, c : 0.375, ab : 0.250, ac : 0.250, bc : 0.250, abc : 0.125\}$ we sort S into $S' = \{abc : 0.125, ab : 0.250, ac : 0.250, bc : 0.250, a : 0.500, b : 0.500, c : 0.375\}$. Starting with the longest element abc its marginal support is equal to its support since there is no superset of abc, thus marginal support of abc is 0.125. Repeating this process for each itemset in S' we get $M = \{abc : 0.125, ab : 0.125, ac : 0.125, bc : 0.125, a : 0.125, b : 0.125, c : 0.000\}$.

3.2   Global Closure

We introduce a new concept, global closure, which is used to determine if a set of itemsets and their supports can efficiently map to a dataset.

Definition 2. *A set of itemsets S has an efficient mapping to a dataset D if it is globally closed. S is globally closed if $\Sigma_{i=1}^{m}(M_i) \leq 1$, that is the sum of all marginal supports are less than or equal to the total support of a dataset, which is 1.*

Itemset	Support	Marginal Support	Marginal Formula		a	b	c	|D| = 8
abc	0.125	0.125	s(abc)		1	1	1	+1 abc
bc	0.250	0.125	s(bc) - m(abc)		0	1	1	+1 bc
ac	0.250	0.125	s(ac) - m(abc)		1	0	1	+1 ac
ab	0.250	0.125	s(ab) - m(abc)		1	1	0	+1 ab
c	0.375	0.000	s(c) - m(ac) - m(bc) - m(abc)		0	1	0	+1 b
b	0.500	0.125	s(b) - m(ab) - m(bc) - m(abc)		1	0	0	+1 a
a	0.500	0.125	s(a) - m(ab) - m(ac) - m(abc)		0	0	0	+2 {}
{}	1.000	0.250	s({}) - m(abc) - m(bc) - m(ac) - m(ab) - m(c) - m(b) - m(a)		0	0	0	

Fig. 3. A globally closed set of itemsets. Arrows indicate the mapping between marginal support and the number of rows in the dataset.

Mapping M onto D. Assuming M is globally closed, we can construct a dataset D of size $|D|$, which is user defined. The algorithm to map M onto D is as follows, sort M from the longest itemset to the shortest, resulting in M'. For each marginal support M'_i that corresponds to an itemset I_i, append the itemset as a new transaction in D such that there are $M'_i \times |D|$ new transactions that contain I_i. Finally, append the marginal support of the empty set to D.

Another interesting property of global closure is the ease of mining a globally closed set of itemsets. Since the marginal supports in M including the empty set are 1, each itemsets support in M has a corresponding amount of support in D as can be seen in Fig. 3.

3.3   Items2Data

Items2Data is an algorithm that enables a user to design a dataset by pre-defining itemsets and their supports. If the pre-defined itemsets are globally closed, then it can guarantee that all supports in the itemsets are satisfied.

Figure 4 is an overview of the Items2Data algorithm. It consists of 3 steps, the first step is to calculate the marginal supports M from a set of itemset supports S which we call Support2Marginal. The second component is to determine whether the marginal supports are globally closed. If M is not globally closed then we do not move onto the third step. If M is globally closed then we move onto the third step, Marginal2Data, which reconstructs a dataset D from the M.

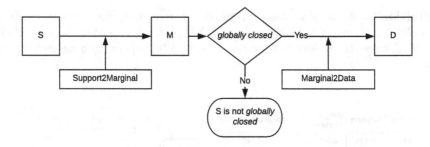

Fig. 4. Items2Data process overview

The benefit of Items2Data is that it is easy to reconstruct desired joint distributions if we can specify a globally closed set of itemsets. For the purposes of designing synthetic data with few items, this is very useful to guarantee that the joint distributions between items will be present. Another benefit is that we have found a tractable solution since the worst case time complexity of Items2Data is determined by the time complexity of the Support2Marginal step which is $O(n^2)$. Although Items2Data works well under the condition where a set of itemsets has to be globally closed, we discuss in future work that this is a first step towards solving the more general problem.

4   Experiments and Results

The experiments in this section were implemented in Python 3.6 using an Intel Core i7-7700HQ processor on a single core. Default algorithms and parameters for Naive Bayes and Decision Trees were used from the scikit-learn library in Python [15]. For each experiment we describe the experimental setup and results.

4.1   Designing Synthetic Data for Classification

It is a common task in machine learning to design synthetic data in order to validate the performance of algorithms under different data distributions. The purpose of this experiment is to test how well Items2Data can generate synthetic data distributions for classification. We utilize the properties of two classifiers, a gaussian based Naive Bayes Classifier (NB) which is known to rely on the conditional independence assumption [16] that does not learn from data distributions which contain conditional dependencies and the Decision Tree Classifier (DT) which does learn from conditional dependencies.

Revisiting the example in Table 1, using itemset a as the target class, Scenario 1 is a synthetic environment where b and c are conditionally independent given a since $p(b|a)p(c|a) = p(b, c|a)$ or equivalently $\frac{ab}{a}\frac{ac}{a} = \frac{abc}{a}$. In contrast, b and c are not conditionally independent in the presence of a for Scenario 2. Under these conditions, if Items2Data has reconstructed the distribution correctly, we would expect similar performance between NB and DT for Scenario 1, whereas the performance of DT will increase in Scenario 2 due to the extra information present in the higher order interaction between variables.

Therefore the experiment is set up as follows, for each scenario, use Items2Data to construct a synthetic dataset. Set the itemset a as the target class with b and c as the features to learn from. Evaluate with NB and DT using 10×10-fold cross validation, then report the average accuracy score and the 95% confidence interval. Reconstruction Accuracy is calculated by $\frac{1}{|S|}\Sigma_{i=1}^{|S|}\frac{R_i}{S_i}$ where S are the pre-defined itemsets with supports and R are the reconstructed itemsets with supports (Table 2).

Table 2. Classification accuracy of Naive Bayes (NB) and Decision Tree (DT) with 2 synthetic datasets

Dataset	Scenario 1	Scenario 2
NB Accuracy	60.0 ± 0.5 (%)	60.0 ± 0.4 (%)
DT Accuracy	60.0 ± 0.5 (%)	**70.0 ± 0.4 (%)**
Reconstruction Accuracy	100%	100%

The results confirm that Items2Data can reconstruct the itemset properties specified in Scenario 1 and 2. Items2Data has also generated a synthetic data distribution in Scenario 1 where DT has similar performance to NB as expected. In Scenario 2 DT, unlike NB, is able to learn the conditional dependence between b and c. Therefore we have shown an example where Items2Data can construct data distributions for machine learning which exhibit specific conditional dependence properties.

4.2 Evaluating Items2Data with Real World Data Characteristics

In this section we demonstrate how Items2Data will perform if we design itemsets that have similar size and characteristics as real world datasets. We use 6 popular frequent itemset datasets from the UCI machine learning repository and other sources [17,18]. For each dataset we calculate 8 properties seen in Table 3. #Transactions is the total number of transactions in a dataset, #Unique Itemsets is the total number unique itemsets, %Unique Itemsets is the percentage of unique itemsets relative to the total number of transactions, #Items is the total number of items and Average Itemset Length is the average of length of all itemsets. Support2Marginal Time is calculated by first deriving the support for all unique itemsets in the dataset which we use to define S. We record the time taken in seconds to construct the marginal supports M from itemset supports S. Marginal2Data Time is the time taken in seconds to reconstruct a dataset D from the marginal supports M with the same size $|D|$ as the original dataset. Reconstruction Accuracy is the same as in the previous experiment.

Table 3 shows the total run time performance of two steps in Items2Data. Support2Marginal scales in polynomial time with #Unique Itemsets as can be

Table 3. Dataset characteristics and reconstruction times

Properties	Accidents	bms1	bms2	Chess	connect4	pumsb
#Transactions	340,183	59,602	77,512	3,196	67,557	49,046
#Unique Itemsets	339,898	18,473	48,684	3,196	67,557	48,474
%Unique Itemsets	0.9992	0.3099	0.6281	1.000	1.000	0.9883
#Items	468	497	3,340	75	129	2,113
Average Itemset Length	33.81	2.51	4.62	37.00	43.00	74.00
Support2Marginal Time	NA	22 s	183 s	1 s	586 s	263 s
Marginal2Data Time	NA	25 s	70 s	1 s	4 s	24 s
Reconstruction Accuracy	NA	100%	100%	100%	100%	100%

seen in Fig. 5. #Unique Itemsets does not completely explain run time performance as it also depends on the average itemset length, which is related to the average number of superset operations required in calculating the marginal support. It can be seen that pumsb has an average itemset length of 74.0 which indicates that there 74 items in each transaction. In non-transaction datasets that have been coerced into a transaction format such as chess and connect4, it is common to see a whole number average itemset lengths. This leads to lower run time since there are no superset subtractions to perform in the marginal support calculation. The lower run time can be seen when comparing Support2Marginal in pumsb to bms2 where they have similar #Unique Itemsets, but pumsb is faster.

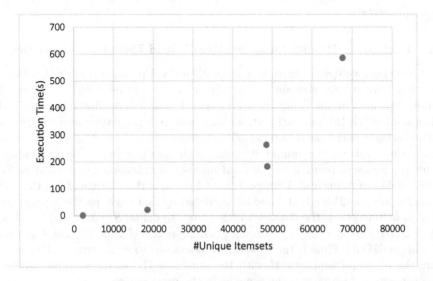

Fig. 5. Execution time in seconds for Support2Marginal by #Unique Itemsets

The accidents dataset was not able to complete due to the number of itemsets being too large. Marginal2Data Time confirms computing a dataset from marginal supports is quick and seems to scale well with #Unique Itemsets. The Reconstruction Accuracy shows that Items2Data can exactly reconstruct the itemset support constraints for the 5 datasets that were able to complete.

5 Conclusion and Future Work

We have proposed an efficient algorithm Items2Data which solves the typically intractable problem of generating a dataset D from a set of itemsets in S by exploiting a condition when S has an efficient mapping to D. We introduce the concept of marginal support and global closure and show that it helps identify when the marginal supports can efficiently map onto D. We also note that a globally closed S can be a more condensed representation of itemsets than using closed itemsets. Lastly we demonstrated the practical application of Items2Data on designing boolean synthetic datasets for experimentation.

Reconstructing data from itemsets efficiently, opens up a new opportunities to generate synthetic boolean data. While the initial approach is a first step and depends on S being globally closed, future work in the ability to repair S so that it is globally closed is important. It potentially shifts the difficulty of IFM from generating data from itemsets to generating globally closed itemsets which if can be addressed efficiently, will lead to many feasible approaches to synthetic boolean data generation. Overall our work has made progress in advancing the intractability and practical application of using itemsets to generate synthetic boolean datasets.

References

1. Trefethen, L., Bau, D.: Numerical Linear Algebra. Other Titles in Applied Mathematics. Society for Industrial and Applied Mathematics (1997)
2. Belohlavek, R., Vychodil, V.: Discovery of optimal factors in binary data via a novel method of matrix decomposition. J. Comput. Syst. Sci. **76**(1), 3–20 (2010)
3. Guzzo, A., Moccia, L., Saccà, D., Serra, E.: Solving inverse frequent itemset mining with infrequency constraints via large-scale linear programs. ACM Trans. Knowl. Disc. Data (TKDD) **7**(4), 18 (2013)
4. Guzzo, A., Saccà, D., Serra, E.: An effective approach to inverse frequent set mining. In: Ninth IEEE International Conference on Data Mining, ICDM 2009, pp. 806–811. IEEE (2009)
5. Wu, X., Wu, Y., Wang, Y., Li, Y.: Privacy-aware market basket data set generation: a feasible approach for inverse frequent set mining. In: Proceedings of the 2005 SIAM International Conference on Data Mining, pp. 103–114. SIAM (2005)
6. Ramesh, G., Zaki, M.J., Maniatty, W.A.: Distribution-based synthetic database generation techniques for itemset mining. In: 9th International Database Engineering and Application Symposium, IDEAS 2005, pp. 307–316. IEEE (2005)
7. Calders, T.: The complexity of satisfying constraints on databases of transactions. Acta Informatica **44**(7–8), 591–624 (2007)

8. Calders, T.: Computational complexity of itemset frequency satisfiability. In: Proceedings of the Twenty-Third ACM SIGMOD-SIGACT-SIGART Symposium on Principles of Database Systems, pp. 143–154. ACM (2004)

9. Pasquier, N., Bastide, Y., Taouil, R., Lakhal, L.: Discovering frequent closed itemsets for association rules. In: Beeri, C., Buneman, P. (eds.) ICDT 1999. LNCS, vol. 1540, pp. 398–416. Springer, Heidelberg (1999). https://doi.org/10.1007/3-540-49257-7_25

10. Mielikainen, T.: On inverse frequent set mining. In: Proceedings of the 3rd IEEE ICDM Workshop on Privacy Preserving Data Mining, pp. 18–23. Citeseer (2003)

11. Madsen, L., Birkes, D.: Simulating dependent discrete data. J. Stat. Comput. Simul. **83**(4), 677–691 (2013)

12. Agrawal, R., Srikant, R., et al.: Fast algorithms for mining association rules. In: Proceedings of the 20th International Conference on Very Large Data Bases, VLDB, vol. 1215, pp. 487–499 (1994)

13. Calders, T., Rigotti, C., Boulicaut, J.-F.: A survey on condensed representations for frequent sets. In: Boulicaut, J.-F., De Raedt, L., Mannila, H. (eds.) Constraint-Based Mining and Inductive Databases. LNCS (LNAI), vol. 3848, pp. 64–80. Springer, Heidelberg (2006). https://doi.org/10.1007/11615576_4

14. Calders, T., Goethals, B.: Non-derivable itemset mining. Data Min. Knowl. Disc. **14**(1), 171–206 (2007)

15. Pedregosa, F., et al.: Scikit-learn: machine learning in Python. J. Mach. Learn. Res. **12**, 2825–2830 (2011)

16. Rish, I.: An empirical study of the Naive Bayes Classifier. In: IJCAI 2001 Workshop on Empirical Methods in Artificial Intelligence, vol. 3, pp. 41–46. IBM (2001)

17. Dheeru, D., Karra Taniskidou, E.: UCI machine learning repository (2017)

18. Geurts, K., Wets, G., Brijs, T., Vanhoof, K.: Profiling of high-frequency accident locations by use of association rules. Transp. Res. Rec. J. Transp. Res. Board **1840**, 123–130 (2003)

Real Time Transaction Management in Replicated DRTDBS

Pratik Shrivastava[(✉)] and Udai Shanker

M.M.M.U.T., Gorakhpur, India
pratik.shrivastav10@gmail.com, udaigkp@gmail.com

Abstract. Real time transaction (RTT) management poses a new challenge in the design of replicated distributed real time database system (RDRTDBS). Existing replication protocols maintain only the mutual consistency between replicated data objects and lack to support the RTT management. This paper explores different scenarios of interaction between coordinator, cohorts & updaters and proposes measures to be taken for each scenario such that performances of the system get improved and at the same time mutual consistency can also be maintained. Additionally, a strict consistency criterion has also been followed to prevent the user from accessing the inconsistent value. This proposed work increases performance in terms of availability, scalability and reliability with respect to other replication protocols.

Keywords: RTT · RDRTDBS · Sub-transaction ·
Dependency relationship · Performance · Mutual consistency

1 Introduction

During past few decades, data is becoming a vital resource for many applications requiring an effective and efficient data management technique [1]. The database system (DBS) is used to store these data such that database operations in terms of searching, insertion, deletion, and updation become easy [2, 3]. DBS can be categorized into two types: centralized and distributed database. In Centralized database system architecture, database operations are executed at a single site that offers more reliability, less overhead and a single point of control whereas, in distributed database system architecture, the database is hosted at diversified locations that are interconnected through an internet/intranet [4, 5].

Real time system (RTS) can be termed as time constrained system whose correctness depends upon the logical consistency of the result and also at the time it is produced [6, 7]. At present, RTS covers a wide spectrum of applications from a simple to very complex one. For example, nuclear power plant, flight control system, space shuttle, and so on. As RTS continues to evolve, many applications are requiring a massive amount of data to be handled in a timely manner [8]. A distributed real time database system (DRTDBS) is specifically designed to handle these data in a timely manner. The primary focus of DRTDBS is on timely completion of real time transaction (RTT) irrespective of logical consistency. Typically, DRTDBS involves distributed execution of RTT and strict consistency requirement that makes to satisfy

© Springer Nature Switzerland AG 2019
L. Chang et al. (Eds.): ADC 2019, LNCS 11393, pp. 91–103, 2019.
https://doi.org/10.1007/978-3-030-12079-5_7

real-time constraint more challenging. Therefore, DRTDBS is usually equipped with a replication technique [9] to address such real-time requirement. The Replicated version of DRTDBS is termed as replicated DRTDBS (RDRTDBS).

In RDRTDBS, data copies are replicated at multiple sites, so that, RTT can be executed locally and performance of the system in terms of availability, scalability, fault-tolerance and reliability can be increased [9]. In order to achieve such advantages, it is necessary that replicated data object should be in consistent state, so that, consistent result can be returned to the user. Therefore, in RDRTDBS majority of research is conducted on the development of efficient and effective replication protocol. Existing replication protocols [10–24] were mainly working to maintain the mutual consistency of the replicated data object. In addition to this, these replication protocols were following different correctness criteria, so that, strict consistency or weaker than strict consistency can be satisfied [25].

In replicated real-time environment, data copies are fully replicated, partially replicated or not replicated. In the current paper, our system model is partially replicated where RTT is admitted on any site. Based on its data requirement, cohorts are established. Likewise, each cohort is having a list of updaters that holds the same data copies. Therefore, for RTT processing, master establishes cohorts and cohort establishes its updaters.

Maintaining mutual consistency in the partially replicated real-time environment is more challenging in comparison to the non-replicated environment because in replicated real-time environment, RTT cannot be committed unless all cohorts and its updaters are committed. Therefore, this paper proposes to use the concept of dependency relationship [26, 27] to improve the performance of RDRTDBS. Adaptively coordinate dependency relationship between sub-transactions executing on coordinator, cohorts, and updaters has been proposed which plays an important role for its performance improvement. This relationship includes the property of symmetric, transitive and likewise. Some dependencies can also involve another dependency. However, extending such concept in a replicated real-time environment is not so easy because of the various factors such as involvement of a large number of sites (i.e. cohorts and their updaters), random arrival of RTT on any site requesting to access the conflicted data object, strict consistency criteria and lack of time. Therefore, to configure the dependency relationship in RDRTBDS, new concept of working set has been proposed which is the set of commit dependency set (CDS), abort dependency set (ADS), termination dependency set (TDS), exclusion dependency set (EDS) and serial dependency set (SDS). In our system, this working set is held by all sites that are executing the sub-transaction of parent RTT. Through this working set, conflicted RTT working on same data item on a particular site can be identified and appropriate resolution mechanism can be applied to further improve the performance of the system.

Although dependency relationship with working set improves the performance of the system, conflicted low priority RTT always has to get blocked or have to wait that may cause low priority RTT to miss its deadline due to the presence of the single version of the data object. Therefore, the concept of dual version data object has been used to overcome such issue. The dual version data object is holding two values i.e. before value and after value. This concept has been already proposed in [10] and has been use in our system. In this paper, a dual version of data object exists on all sites such that more than one RTT can work on the same data object. Here, parallel update

RTT and read RTT can execute on the same real-time data object. Similarly, write RTT and read RTT can execute on non-real time data object. This parallel execution prevents the wastage of resources because if unconditionally high priority RTT gets aborted then parallel low priority RTT gets the chance to commit.

Overall, our proposed concept of dependency relationship in the replicated real time environment coordinates different dependency relationship between sub transactions executing on different sites, uses working set to identify the conflicted RTT executing on the same site, adapts proper mechanism to further improve the performance of the system and uses dual version data object that avoids wastage of resources such that resources can be properly utilized.

The main contribution of this paper is three-fold.

1. Specifies and coordinates different dependency relationships between different sub-transactions to further improve the performance in terms of transaction miss ratio.
2. Conflict detection and resolution are done with the help of our new concept of working set.
3. Utilization of dual version of data objects in place of single version such that wastage of resources can be prevented.

The rest of the paper is organized as follows. Section 2 introduces the dependency relationship. Section 3 discusses system model and presents the new concept of working set for the dependency relationship. In Sect. 4, a mechanism to improve performance of the system has been proposed. Section 5 discusses the simulation results and Sect. 6, finally, concludes the paper.

2 Dependency Relationship

The Dependency relationship is used to specify and coordinate different dependency relationship between sub-transactions executing on coordinator, cohorts and updaters. The decision of relationship is identified from the read and write set of sub-transactions. In our system, a dependency relationship is linked with an appropriate action that defines what is to be done so that performance can be improved. For instance, during strong commit dependency if sub-transaction in cohort gets committed, all its updaters must also commit. Similarly, during abort dependency, if sub-transaction in the cohort is aborted, all its updaters must also get aborted.

Dependency relationship can be categorized into external dependency, internal dependency and independent dependency. The brief description of each dependency is given as follows.

2.1 External Dependency

This dependency can be sub-categorized into exclusion and termination dependency. Its definition is as follows.

Exclusion Dependency

In this dependency, the outcome of one sub-transaction decides the outcome of another sub-transaction. For example, in between two sub-transaction TI and TJ, if TI commits then TJ must abort. It is represented as TI-*E*-TJ.

Termination Dependency

In this type of dependency, sub-transaction executing cannot unilaterally commit or abort until the coordinator decides about commit or abort. Similarly, updaters are also not allowed to commit or abort until cohorts decide to commit or abort. It is represented as Coordinator-*T*-Cohorts and Cohorts-*T*-Updaters.

2.2 Internal Dependency

This dependency can be sub-categorized into strong commit and weak abort dependency. Its description is as follows.

Strong Commit Dependency

This dependency exists between the coordinator & its cohorts and in between cohort & its updaters. Action attached to this relationship defines that in between two sub-transactions TI and TJ if TI is committed then TJ must also be committed. It is represented as TI-*SC*-TJ.

Weak Abort Dependency

This dependency also exists between the coordinator and its cohorts, and in between cohort and its updaters. Action linked with this dependency directs that in between TI and TJ if TI is aborted then TJ will also get aborted. It is represented as TI-*A*-TJ.

2.3 Independent Dependency

As already defined in the previous paragraph, an independent dependency exists in between those sub-transactions that do not use common data objects for their execution. It is represented as TI-*I*-TJ.

Now, consider a real-time scenario that makes a clear-cut understanding of how dependency relationship exists.

Example 2.1. Suppose, there exist three transactions T1, T2, T3 that consist of a list of operations and their operands (i.e. data objects are V, X, Y, and Z). These all transactions are submitted to a node to be completed within their deadline. The detailed information in the form of a group of operations is given below.

T1:R1(X), W1(X), R1(Y), W1(Y)
T2:R2(Z), W2(Z)
T3:R3(V), W3(V), R3(Y), W3(Y)

As given in Fig. 1. T1 is submitted at node N4 for execution. The required data objects for T1 is X and Y. Here, N4 holds the copy of X. So, it can execute the operation related to X only. For executing the operation related to X, sub transaction T11 is created on node N4. In addition to this, N3 and N5 are the updaters of N4. To maintain the consistency of data object X, N4 requests for LOCK on N3 and N5 also, such that, forthcoming transaction for example transactions on-site N3 or N5 cannot alter the

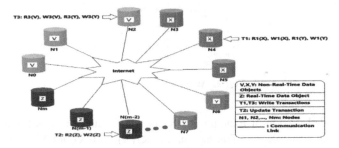

Fig. 1. Snapshot of RDRTDBS

*value of X. To improve the concurrency, we use here a dual version of X, such that, transaction does not have to wait and can use the before value of data object to proceed. Therefore, this transaction is exclusive dependent on T11 (i.e. T11-E-TXN). In general, N3 and N5 are related to N4 through strongly committed dependency, abort dependency and termination dependency (i.e. N4-**SC**-N3, N4-**SC**-N5, N4-**A**-N3, N4-**A**-N5, N4-**T**-N3, and N4-**T**-N5). To maintain the consistency at N3 and N5, sub-transaction T11_1 and T11_2 will be executed at N3 and N5 respectively. Similarly, to execute the operation on Y, cohort N6 is initiated. To complete the execution on N6, sub-transaction T12 is created. As shown in Fig. 1, N7 is linked as an updater of N6. Therefore, here N6 and N7 are also related with strongly committed dependency, abort dependency and termination dependency (i.e. N6-**SC**-N7, N6-**A**-N7, and N6-**T**-N7). To maintain the consistency in between cohort(N6) and updater(N7), N7 will execute the transaction T12_1. In addition to this, N4 and N6 are also related with strongly committed dependency & abort dependency (i.e. N4-**SC**-N6 and N4-**A**-N6).*

3   System Model

Before we describe proposed algorithm, our main objective is to first introduce about RDRTDBS system model. Our assumptions with respect to locking model, network model, database model and other parameters are same as given in [14]. Additionally, each sub-transaction is connected with a working set that consists of different subsets (i.e. CDS, ADS, TDS, EDS, and SDS). In the shown Fig. 2, CDS stands for commit dependency subset which stores the id of those sub-transactions that have to get committed. Similarly, ADS stand for abort dependency subset which holds the id of those sub-transactions that must get aborted. Likewise, EDS is an abbreviation for exclusion dependency subset which holds the id of those sub-transactions that are exclusively dependent. TDS stands for termination dependency subset; this set holds the id of those sub-transactions that cannot unilaterally decide to commit or abort. Finally, SDS is the last set, which holds the id of those sub-transactions that are waiting to execute.

Parent Transaction-id
CDS = {Commit dependent sub-transactions}
ADS = {Abort dependent sub-transactions}
TDS = {Terminate dependent sub-transactions}
EDS = {Exclusion dependent sub-transactions}
SDS = {Serially dependent sub-transactions}

Fig. 2. Working set for RTT.

The extended version of Example 2.1 using the working set is as follows.

As given in Example 2.1, Transaction T1: R1(X), W1(X), R1(Y), W1(Y) is submitted for execution at N4. To complete the transaction T1, sub-transaction T11 will be created at N4 and T12 will create cohort at N6. Here, N3 and N5 are acting as updaters for N4 and N7 is acting as an updater for N6. Therefore, to maintain the consistency between N3, N4 & N5, sub-transaction T11_1 and T11_2 will be created at N3 and N5 respectively. Likewise, T12 will execute at N6 and to maintain the consistency between N6 and N7, sub-transaction T12_1 is created at N7. Overall, five working set will be created for T1 (i.e. T11, T11_1, T11_2, T12, and T12_1). The working set for T11 is given in Fig. 3. In this working set. CDS contains the id of sub-transactions T11_1, T11_2, and T12 because, if T11 get commit then T11_1, T11_2 and T12 will have to commit. Similarly, ADS also hold the id of sub-transactions T11_1, T11_2, and T12. In the working set of T11, EDS and SDS are empty because there does not exist any sub-transaction which are exclusively and serially dependent.

Transaction-Id (T11)
CDS = {T11_1, T11_2, T12}
ADS = {T11_1, T11_2, T12}
TDS = {T12}
EDS = {}
SDS = {}

Fig. 3. Working set for Transaction (T11).

Let read RTT "Transaction" is admitted at N3 to read X value. However, transaction gets conflicted with already executing T11_1 because both sub-transactions are accessing common data object X. Therefore, transaction will use before value of X to gets executed without wait and T11_1 will use after value to update X. So, in the working set of T11_1, CDS and ADS contain the id of sub-transactions T11_1 and EDS hold the id of sub-transaction transaction. The working set of T11_1 is shown in the Fig. 4.

Transaction-Id (T11_1)
CDS = {T11_1}
ADS = {T11_1}
TDS = {}
EDS = {TXN}
SDS = {}

Fig. 4. Working set for Transaction (T11_1).

Figure 5 is the working set of T11_2 and is presented at N5. In this working set, CDS and ADS hold the id of T11_2 because at N5, there is no any other transaction to execute. Similarly, TDS, EDS, SDS subsets are empty because there does not exist any other transaction which is exclusively, terminate and serially dependent.

Transaction-id (T11_2)
CDS = {T11_2}   ADS = {T11_2}   TDS = {}   EDS = {}   SDS = {}

Fig. 5. Working set for Transaction (T11_2).

Figure 6 is the working set of T12 and is present at N6. In this working set, CDS and ADS hold the id of T12, whereas, TDS hold the id of T31. T31 is the sub-transaction created at N6 to access the data object Y.

Transaction-id (T12)
CDS = {T12}   ADS = {T12}   TDS = {T31}   EDS = {}   SDS = {}

Fig. 6. Working set for Transaction (T12).

In the working set of T12_1, CDS and ADS hold the id of T12_1 whereas TDS holds the id of T31_1. Here, T31_1 is acting as updater for T31. Overall, T1 will complete its installation by establishing the sub-transactions T11, T11_1, T11_2, T12, and T12_1. Figure 7 is the working set of T12_1.

Transaction-id (T12_1)
CDS = {T12_1}   ADS = {T12_1}   TDS = {T31_1}   EDS = {}   SDS = {}

Fig. 7. Working set for Transaction (T12_1).

4 Mechanism of Interaction Between Coordinator, Cohort and Updaters

In the previous section, we have briefly explained about dependency relationship and working set through an example. In this section, our intention is to introduce different scenarios and propose proper actions, so that, performance can be improved. Four scenarios have been considered wherein each scenario proper actions are specified that are initiated automatically based on the sub-transaction status.

4.1 Scenario 1: T11 Decides to Commit Before T5 Completed Its Local Data Processing

On Coordinator Site

If T11 decides to commit.

> *T11 will send commit message to cohorts present in CDS.*
> *Wait for a response from all cohorts.*

If the global decision is to commit.

> *T5 will enqueue on the wait queue, T11 will get committed, and EDS, TDS and ADS subset of T11 get discarded.*

If the global decision is to abort.

> *T5 will get the opportunity to get a commit, and EDS, TDS, ADS of T5 get discarded.*

On Cohort Site

If T12 decides to commit.

> *T12 will send commit message to T12_1 to get a commit.*
> *Wait for a response from updaters.*

If the global response is to commit from all updaters.

> *T12 get committed and reply with commit message to T11. EDS, TDS, and ADS of T12 get discarded.*

If the global message is to abort from all updaters.

> *T12 get abort and reply with abort message to T11. EDS, TDS, and CDS of T12 get discarded.*

4.2 Scenario 2: T5 Decides to Commit Before T11 Completed Its Local Data Processing

On Coordinator Site

If T5 decides to commit.

> *T11 will send commit message to cohorts present in CDS.*
> *Wait for a response from all cohorts.*

If the global decision is to commit.

> *T11 will enqueue on the wait queue, T5 will get committed, and EDS, TDS and ADS subset of T5 get discarded.*

If the global decision is to abort.

> *T11 will get the opportunity to get a commit, and EDS, TDS, ADS of T11 get discarded.*

On Cohort Site
If T5 decides to commit.

> *T5 will send commit message to its updaters.*
> *Wait for a response from updaters.*

If the global response is to commit from all updaters.

> *T5 get a commit and EDS, TDS, and ADS of T5 get discarded.*

If the global message is to abort from all updaters.

> *T5 get abort and EDS, TDS and CDS of T12 get discarded.*

4.3 Scenario 3: T5 and T11 Both Decide to Commit at the Same Time

On Coordinator Site
High priority transaction gets the chance to get committed.
If high priority transaction gets committed, then low priority transaction gets restarted.
If unconditionally high priority transaction gets aborted, then waiting for lower priority transaction will get the chance to get committed.

4.4 Scenario 3: T5 and T11 Both Decide to Abort at the Same Time

On Coordinator Site
Sub-transaction waiting on SDS of T5 and T11 gets a chance to proceed.

5 Simulation and Experimental Result

To check the performance of our proposed mechanism, we develop a detailed simulation model for RDRTDBS. Our model is based on the model presented in [14].

5.1 Simulation Setting

The main objective of developing simulator and performing experiments is to get ensured that proposed algorithms improve the performance in terms of transaction miss ratio. In addition to this, our intention is to perform experiments with changing parameters, condition and their results are explained in this section. A list of parameters and their respected values used in the simulation model are presented in Table 1.

Table 1. User transaction workload and system parameter setting

Parameter	Values
Number of nodes	10
Number of pages in the database	1000 pages
Replication Degree	4
CPUs per node	2
Data disks per node	4
Log disks per node	1
Buffer hit ratio on a node	0.1
Transaction category	Sequential
Arrival rate (Trans./Second)	Varies
Slack value	6.0
No. of pages accessed per transaction	10 pages

Deadline assignment for the transaction is calculated by the formula given below.

$$DL = AT + ST * Rt \tag{1}$$

Where DL, AT, SL and Rt are the deadline, arrival time, slack factor value and resource time respectively of the requested transaction.

5.2   Experimental Result

We compare the performance of our model with another existing replication protocol [14]. The experimental results have been presented where each simulation run 10 times and the majority of times (i.e. 92% times) confidence intervals are drawn for each data point. Figures 8 and 9 show improvement in missed deadline percentage when compared with MIRROR under normal and heavy load, Fig. 10 shows the improvement in

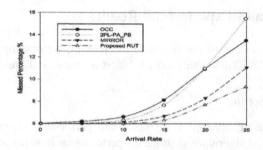

Fig. 8. Miss % under normal load.

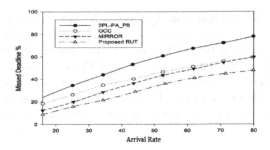

Fig. 9. Miss % under heavy load.

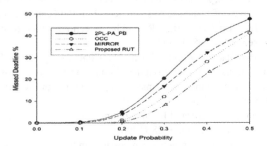

Fig. 10. Miss % under low update frequency.

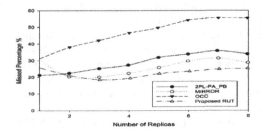

Fig. 11. Miss % with partial replication.

missed deadline under low update frequency and Fig. 11 shows improved missed deadline percent with partial replication.

6 Conclusion

In this paper, our objective was to further improve the performance of RDRTDBS. This objective was completed via coordinating different dependency relationship between sub-transactions and by executing proper actions. In addition to this, we have proposed a new concept of working set that identifies the conflicted sub-transactions executing on a site and gives the opportunity to high priority RTTs to get completed. To prevent the wastage of the system resources, we have proposed to use the dual version of data object that allows parallelly two RTT to work on the common data object. Additionally, we have explored the different scenario of RTT processing and have proposed proper measures for each scenario. The performance of our proposed dependency relationship was compared with the MIRROR and simulations result shows that our proposed RUT improves the performance in terms of transaction miss ratio.

References

1. Berrington, J.: Databases. Anaesth. Intensive Care Med. **8**(12), 513–515 (2017)
2. Ramakrishnan, R., Gehrke, J.: Database Management Systems. McGraw Hill, Boston (2000)
3. Ullman, J.D.: Principles of Database Systems. Galgotia Publications, New Delhi (1984)
4. Garcia-Molina, H., Lindsay, B.: Research directions for distributed databases. ACM SIGMOD Rec. **19**(4), 98–103 (1990)
5. Bernstein, P.A., Hadzilacos, V., Goodman, N.: Concurrency Control and Recovery in Database Systems. Addison-Wesley Longman Publishing Co., Inc., Boston (1987)
6. Aldarmi, S.A.: Real-time database systems: concepts and design. Report-University of York Department of Computer Science YCS (1998)
7. Thomas, S., Seshadri, S., Haritsa, J.R.: Integrating standard transactions in firm real-time database systems. Inf. Syst. **21**(1), 3–28 (1996)
8. Ulusoy, O.: A study of two transaction processing architectures for distributed real-time database systems. J. Syst. Softw. **31**(2), 97–108 (1995)
9. Shrivastava, P., Shanker, U.: Replica update technique in RDRTDBS: issues & challenges. In: ADCOM-2018 Ph.D. Forum, August 2018 (Accepted)
10. Son, S.H., Kouloumbis, S.: A token-based synchronization scheme for distributed real-time databases. Inf. Syst. **18**(6), 375–389 (1993)
11. Son, S.H., Zhang, F.: Real-time replication control for distributed database systems: algorithms and their performance. In: DASFAA, vol. 11, pp. 214–221, April 1995
12. Son, S.H., Zhang, F., Hwang, B.: Concurrency control for replicated data in distributed real-time systems. J. Database Manage. (JDM) **7**(2), 12–23 (1996)
13. Kim, Y.K.: Towards real-time performance in a scalable, continuously available telecom DBMS (1996)
14. Xiong, M., Ramamritham, K., Haritsa, J.R., Stankovic, J.A.: MIRROR: a state-conscious concurrency control protocol for replicated real-time databases. Inf. Syst. **27**(4), 277–297 (2002)
15. Peddi, P., DiPippo, L.C.: A replication strategy for distributed real-time object-oriented databases. In: Fifth IEEE International Symposium on Object-Oriented Real-Time Distributed Computing, ISORC 2002. Proceedings, pp. 129–136. IEEE (2002)
16. Gustavsson, S., Andler, S.F.: Real-time conflict management in replicated databases. In: Proceedings of the Fourth Conference for the Promotion of Research in IT at New Universities and University Colleges in Sweden, PROMOTE IT 2004, Karlstad, Sweden, vol. 2, pp. 504–513 (2004)
17. Gustavsson, S., Andler, S.R.: Continuous consistency management in distributed real-time databases with multiple writers of replicated data. In: 19th IEEE International Parallel and Distributed Processing Symposium, Proceedings, 8 pp. IEEE, April 2005
18. Syberfeldt, S.: Optimistic replication with forward conflict resolution in distributed real-time databases, Doctoral dissertation, Institutionen för datavetenskap (2007)

19. Haj Said, A., Sadeg, B., Amanton, L., Ayeb, B.: A protocol to control replication in distributed real-time database systems. In: Proceedings of the Tenth International Conference on Enterprise Information Systems, ICEIS, vol. 1, pp. 501–504 (2008). ISBN 978-989-8111-36-4

20. Mathiason, G., Andler, S.F., Son, S.H.: Virtual full replication by adaptive segmentation. In: 13th IEEE International Conference on Embedded and Real-Time Computing Systems and Applications, RTCSA 2007, pp. 327–336. IEEE, August 2007

21. Salem, R., Saleh, S.A., Abdul-kader, H.: Scalable data-oriented replication with flexible consistency in real-time data systems. Data Sci. J. **15**, 4 (2016)

22. Shrivastava, P., Shanker, U.: Replica control following 1SR in DRTDBS through best case of transaction execution. In: Kolhe, M., Trivedi, M., Tiwari, S., Singh, V. (eds.) Advances in Data and Information Sciences. LNCS, vol. 38, pp. 139–150. Springer, Singapore (2018). https://doi.org/10.1007/978-981-10-8360-0_13

23. Shrivastava, P., Shanker, U.: Supporting transaction predictability in replicated DRTDBS. In: Proceedings of the 15th International Conference on Distributed Computing and Internet Technology (ICDCIT), Bhubaneshwar, India, 10–13 January 2019 (Accepted)

24. Shrivastava, P., Shanker, U.: Replication protocol based on dynamic versioning of data object for replicated DRTDBS. In: Proceedings of the International Conference on Computational Intelligence & Internet of Things (ICCIIoT), Agartala, India, 14–15 December 2018 (Accepted)

25. Ouzzani, M., Medjahed, B., Elmagarmid, A.K.: Correctness criteria beyond serializability. In: Liu, L., Özsu, M.T. (eds.) Encyclopedia of Database Systems, pp. 501–506. Springer, Boston (2009)

26. Xin, T.: A Framework for Processing Generalized Advanced Transactions. Colorado State University (2006)

27. Shanker, U., Misra, M., Sarje, A.K., Shisondia, R.: Dependency sensitive shadow SWIFT. In: 10th International Database Engineering and Applications Symposium, IDEAS 2006, pp. 273–276. IEEE, December 2006

Materialized View Selection
for Aggregate View Recommendation

Humaira Ehsan[(✉)] and Mohamed A. Sharaf

School of Information Technology and Electrical Engineering,
The University of Queensland, Brisbane, QLD, Australia
{h.ehsan,m.sharaf}@uq.edu.au

Abstract. Data analysts arduously rely on data visualizations for draw-
ing insights into huge and complex datasets. However, finding interesting
visualizations by manually specifying various parameters such as type,
attributes, granularity is a protracted process. Simplification of this process requires systems that can automatically recommend interesting visualizations. Such systems primarily work first by evaluating the utility of
all possible visualizations and then recommending the top-k visualizations to the user. However, this process is achieved at the hands of high
data processing cost. That cost is further aggravated by the presence
of numerical dimensional attributes, as it requires binned aggregations.
Therefore, there is a need of recommendation systems that can facilitate
data exploration tasks with the increased efficiency, without compro-
mising the quality of recommendations. The most expensive operation
while computing the utility of the views is the time spent in executing
the query related to the views. To reduce the cost of this particular operation, we propose a novel technique $mView$, which instead of answering
each query related to a view from scratch, reuses results of the already
executed queries. This is done by incremental materialization of a set of
views in optimal order and answering the queries from the materialized
views instead of the base table. The experimental evaluation shows that
the $mView$ technique can reduce the cost at least by 25–30% as compared
to the previously proposed methods.

1 Introduction

With the unprecedented increase in the volume of data, the challenge of finding
efficient ways to extract interesting insights is critical. As such, data visualization
has become the most common and effective tool for exploring such insights. Generally, the visualizations are generated using user-driven tools like Tableau, Qlik,
Microsoft Excel, etc. However, the use of these tools is of limited effectiveness
for large datasets, as it is very difficult for the user to manually determine the
best data visualization by sequentially browsing through the available representations. Research efforts are therefore being directed to propose recommendation
systems that *automatically recommend visualizations* [3,4,8–10,13]. These systems automatically manipulate the user selected dataset, generate all possible

© Springer Nature Switzerland AG 2019
L. Chang et al. (Eds.): ADC 2019, LNCS 11393, pp. 104–118, 2019.
https://doi.org/10.1007/978-3-030-12079-5_8

visualizations, and recommend the top-k interesting visualizations, where inter-estingness is quantified according to some utility function such as deviation, similarity, diversification, etc. The generation of these all possible visualizations is challenged by a wide range of possible factors. This include user-driven factors such as individual user preferences, data of interest, information semantics and tangible factors such as chart type, possible attribute combinations and available transformations (e.g. sorting, grouping, aggregation and binning).

Recent studies have focused on automatically generating all possible *aggregate views* of data and proposing search strategies for finding the top-k views for recommendation, based on the deviation based utility metric [3,4,13]. The search space of all possible visualizations is huge and it explodes even further in the presence of *numerical dimensions*, as *binned aggregation* is required to group the numerical values along a dimension into adjacent intervals. In our previous work [3,4] on visualization recommendation, binning for numeric dimensions was introduced and efficient schemes (named as *MuVE*) to recommend the top-k binned views were proposed.

The most expensive operation while computing the utility of views is the time spent in executing the queries related to the views. To reduce the cost of this particular operation, a novel technique *mView* is proposed, which instead of answering each query related to a view from scratch, *reuses results* from the already executed queries. In summary, this is done by materializing views and answering queries from the materialized views instead of the base table. The idea of materializing views for reducing the query-processing time is well studied in the literature [2,6,11] and has proven significant relevance to a wide variety of domains, such as query optimization, data integration, mobile computing and data warehouse design [6,11]. However due to prohibitively large number of views, the blind application of materialization may result in even further degradation of the cost [2]. Substantial amount of work has already been done to select an appropriate set of views to materialize that minimize the total query response time and the cost of maintaining the selected views, given a limited amount of resource, e.g., materialization time, storage space etc. [5].

In this work our proposed technique *mView* first defines a cost benefit model to decide which views are the best to reuse. Later, in an optimal order it mate-rializes the best set of views, which reduce the overall cost of the solution.

The main contributions of this work are as follows:

- We formulate and analyze the problem of selecting views to materialize for efficiently generating aggregate views in the presence of numerical attribute dimensions (Sect. 3).
- We propose the *mView* technique, which introduces a novel search algorithm, particularly optimized to leverage the specific features of the binned views (Sect. 4).
- We conduct extensive experimental evaluation, which illustrate the benefits achieved by *mView* (Sects. 5.1 and 5.2).

2   Preliminaries

2.1   Aggregate View Recommendation

The process of visual data exploration is typically initiated by an analyst specifying a query Q on a database D_B. The result of Q, denoted as D_Q, represents a subset of the database D_B to be visually analyzed. For instance, consider the following query Q:

Q: SELECT * FROM D_B WHERE T;

In Q, T specifies a combination of predicates, which selects a portion of D_B for visual analysis. A visual representation of Q is basically the process of generating an aggregate view V of its result (i.e., D_Q), which is then plotted using some of the popular visualization methods (e.g., bar charts, scatter plots, etc.). Similar to traditional OLAP systems and recent data visualization platforms [8,9,12,13], our model is based on a multi-dimensional database D_B, consisting of a set of dimension attributes $\mathbb{A}$ and a set of measure attributes $\mathbb{M}$. Additionally, $\mathbb{F}$ is the set of possible aggregate functions over the measure attributes $\mathbb{M}$, such as SUM, COUNT, AVG, STD, VAR, MIN and MAX. Hence, an aggregate view V_i over D_Q is represented by a tuple (A, M, F) where $A \in \mathbb{A}$, $M \in \mathbb{M}$, and $F \in \mathbb{F}$. That is, D_Q is grouped by dimension attribute A and aggregated by function F on measure attribute M. A possible view V_i of the example query Q above would be expressed as:

V_i: SELECT A, F(M) FROM D_B WHERE T GROUP BY A;

where the GROUP BY clause specifies the dimension A for aggregation, and $F(M)$ specifies both the aggregated measure M and the aggregate function F.

Typically, a data analyst is keen to find visualizations that reveal some interesting insights about the analyzed data D_Q. However, the complexity of this task stems from: (1) the large number of possible visualizations, and (2) the interestingness of a visualization is rather subjective. Towards automated visual data exploration, recent approaches have been proposed for recommending interesting visualizations based on some objective, well-defined quantitative metrics (e.g., [8,9,13]). Among those metrics, recent case studies have shown that a *deviation-based* metric is able to provide interesting visualizations that highlight some of the particular trends of the analyzed datasets [13].

In particular, the deviation-based metric measures the distance between $V_i(D_Q)$ and $V_i(D_B)$. That is, it measures the deviation between the aggregate view V_i generated from the subset data D_Q vs. that generated from the entire database D_B, where $V_i(D_Q)$ is denoted as *target* view, whereas $V_i(D_B)$ is denoted as *comparison* view. The premise underlying the deviation-based metric is that a view V_i that results in a higher deviation is expected to reveal some interesting insights that are very particular to the subset D_Q and distinguish it from the general patterns in D_B. To ensure that all views have the same scale, each target view $V_i(D_Q)$ is normalized into a *probability distribution* $P[V_i(D_Q)]$ and each comparison view into $P[V_i(D_B)]$.

For a view V_i, given the probability distributions of its target and comparison views, the deviation $D(V_i)$ is defined as the distance between those probability distributions. Formally, for a given distance function *dist* (e.g., Euclidean distance, Earth Mover's distance, K-L divergence, etc.), $D(V_i)$ is defined as:

$$D(V_i) = dist(P[V_i(D_Q)], P[V_i(D_B)]) \tag{1}$$

Consequently, the deviation $D(V_i)$ of each possible view V_i is computed, and the k views with the highest deviation are recommended (i.e., *top-k*) [13]. Hence, the number of possible views to be constructed is $N = 2 \times |\mathbb{A}| \times |\mathbb{M}| \times |\mathbb{F}|$, which is clearly inefficient for a large multi-dimensional dataset.

2.2 Binned Views

In the previous section we discussed about aggregate view recommendation specifically for categorical dimensions. However, for continuous numerical dimensions, typically the numerical values along a dimension require grouping into adjacent intervals over the range of values. For example, consider a table of employees, which has *Age* as a numerical dimension attribute. Particularly, one of the aggregate views on this attribute is count the number of employees grouped by *Age*. For this type of views, it is more meaningful if adjacent intervals are grouped together and shown in a summarized way. For example, Fig. 1a shows the whole range grouped in 8 bins.

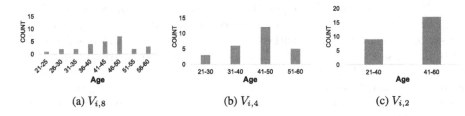

(a) $V_{i,8}$ (b) $V_{i,4}$ (c) $V_{i,2}$

Fig. 1. Generating $V_{i,2}$ by performing aggregation on $V_{i,4}$ or $V_{i,8}$

To enable the incorporation and recommendation of visualizations that are based on numerical dimensions, in our previous work [3,4], we introduced the notion of a *binned view*. A binned view $V_{i,b}$ simply extends the basic definition of a view to specify the applied binning aggregation. Specifically, given a view V_i represented by a tuple (A, M, F), where $A \in \mathbb{A}$, $M \in \mathbb{M}$, $F \in \mathbb{F}$, and A is a continuous numerical dimension with values in the range $L = [L_{min} - L_{max}]$, then a binned view $V_{i,b}$ is defined as:

Binned View: Given a view V_i and a bin width of w, a binned view $V_{i,b}$ is a representation of view V_i, in which the numerical dimension A is partitioned into a number of b equi-width non-overlapping bins, each of width w, where $0 < w \leq L$, and accordingly, $1 \leq b \leq \frac{L}{w}$.

For example, Fig. 1a shows a binned view $V_{i,8}$, in which the number of bins $b = 8$ and the bin width $w = 5$, while Fig. 1c shows a binned view $V_{i,2}$, in which the number of bins $b = 2$ and the bin width $w = 20$. Note that this definition of a binned view resembles that of an equi-width histogram in the sense that a bin size w is uniform across all bins. While other non-uniform histograms representations (e.g., equi-depth and V-optimal) often provide higher accuracy when applied for selectivity estimation, they are clearly not suitable for standard bar chart visualizations. Given our binned view definition, a possible binned bar chart representation of query Q is expressed as:

$$V_{i,b} : \text{SELECT A, F(M) FROM } D_B \text{ WHERE T GROUP BY A}$$
$$\text{NUMBER OF BINS b}$$

The deviation provided by a binned view $V_{i,b}$ is computed similar to that in Eq. 1. In particular, the comparison view is binned using a certain number of bins b and normalized into a probability distribution $P[V_{i,b}(D_B)]$. Similarly, the target view is binned using the same b and normalized into $P[V_{i,b}(D_Q)]$. Then the deviation $D(V_{i,b})$ is calculated as:

$$D(V_{i,b}) = dist(P[V_{i,b}(D_Q)], P[V_{i,b}(D_B)]) \tag{2}$$

2.3   View Processing Cost

Recall that in the absence of numerical dimensions, the number of candidate views N to be constructed is equal to $2 \times N$, where $N = |\mathbb{A}| \times |\mathbb{M}| \times |\mathbb{F}|$. In particular, $|\mathbb{A}| \times |\mathbb{M}| \times |\mathbb{F}|$ queries are posed on the data subset D_Q to create the set of target views, and another $|\mathbb{A}| \times |\mathbb{M}| \times |\mathbb{F}|$ queries are posed on the entire database D_B to create the corresponding set of comparison views. For each candidate non-binned view V_i over a numerical dimension A_j, the number of target and comparison binned views is equal to: $|\mathbb{M}| \times |\mathbb{F}| \times B_j$ each, where B_j is the maximum number of possible bins that can be applied on dimension A_j (i.e., number of binning choices). Hence, in the presence of $|\mathbb{A}|$ numerical dimensions, the total number of binned views grows to N_B which is simply calculated as:

$$N_B = 2 \times \sum_{j=1}^{|\mathbb{A}|} |\mathbb{M}| \times |\mathbb{F}| \times B_j \tag{3}$$

Furthermore, each pair of target and comparison binned views incur query execution time and deviation computation time. Query execution time is the time required to process the raw data to generate the candidate target and comparison binned views, where the cost for generating the target view is denoted as $C_t(V_{i,b})$, and that for generating the comparison view is denoted as $C_c(V_{i,b})$. Moreover, deviation computation time is the time required to measure the deviation between the target and comparison binned views, and is denoted as: $C_d(V_{i,b})$. Notice that this time depends on the employed distance function $dist$.

Putting it together, the total cost incurred in processing a candidate view V_i is expressed as:

$$C(V_i) = \sum_{b=1}^{B} C_t(V_{i,b}) + C_c(V_{i,b}) + C_d(V_{i,b}) \tag{4}$$

We note that the cost of computing deviation is negligible as compared to query execution cost, as it involves no I/O operations. Furthermore, for simplicity in the next sections we assume $C(V_{i,b}) = C_t(V_{i,b}) + C_c(V_{i,b})$. Therefore, Eq. 4 is reduced to:

$$C(V_i) = \sum_{b=1}^{B} C(V_{i,b}) \tag{5}$$

Hence, the total cost incurred in processing all candidate binned views is expressed as:

$$C = \sum_{i=1}^{N} C(V_i) \tag{6}$$

The goal of this study is to propose schemes that reduce the cost $C_t(V_{i,b})$ and $C_c(V_{i,b})$, which will consequently reduce the overall cost C of the solution.

3   Problem: Materialized View Selection

As mentioned in Sect. 2, the view recommendation process involves the generation of a huge number of the comparison and the target views. Particularly, these views are the result of executing their corresponding aggregate queries. Section 2.3 outlines how colossal the cost is for the binned view recommendation problem. However, we notice that for binned aggregate queries, the result of certain queries can be used to answer other queries. For instance, consider a view $V_{i,2} = (A, M, F, 2)$ can be answered from a number of other views such as $V_{i,4}$, $V_{i,6}$, $V_{i,8}$, etc., by performing aggregation on these views instead of the base table. We term this relationship as *dependency*. For instance, view $V_{i,2}$ depends on $V_{i,4}$, $V_{i,6}$ and $V_{i,8}$.

Definition: View Dependency: a binned view $V_{i,b}$ depends on another binned view $V_{i,b'}$, if $V_{i,b}$ can be answered using $V_{i,b'}$, where b' is a multiple of b i.e., $b' = xb$.

For any non-binned view V_i, all the possible binned views $V_{i,b}$ can be directly generated from the base table. Therefore, every $V_{i,b}$ at least depends on the base table, and at most depends on $\frac{B}{b} - 1$ other views $V_{i,b'}$. The dependency relationship between the candidates can be represented by a lattice. Figure 2 shows the lattice for a particular non-binned view V_i that can have a maximum of 8 bins. Each node in the lattice represents a binned view, e.g. node 5 is binned view $V_{i,5}$, while node 0 represents the base table. A view can be generated using any of its ancestors in the lattice. For instance, the ancestors of node 3 (i.e., $V_{i,3}$) are node 6 (i.e., $V_{i,6}$) and node 0 (i.e., base table).

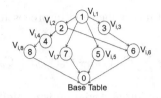

Fig. 2. Lattice for view V_i with $B = 8$

Every $V_{i,b'}$ is a candidate view that can be reused to generate some other views. Specifically, $V_{i,b}$ can be cached in the memory or stored on the disk for later reuse. However, because of the limited memory it is practical to store the view on the disk. Therefore, we propose to materialize the views that are later required to be reused. For instance, in Fig. 2, every view that is an ancestor of at least one other view is a candidate view to be materialized. A key problem is how to decide which views should be reused? The three possible options are:

1. *Reuse nothing:* This is the baseline case in which all the queries are answered from the base table. Consequently, this would incur the query processing time for each binned view from scratch.
2. *Reuse the whole lattice:* In this case all views should be materialized. This would reduce the query processing time of each binned view but the overall execution time of the solution will increase because it would include the additional cost of materializing the views.
3. *Reuse a set of views:* Choose an optimal set of views $\mathbb{T}$ to reuse and materialize them. This will incur the cost of materialization but reduce the overall cost of the solution because a number of queries will be answered from the materialized views instead of the base table.

The best option is to reuse a set of views, which has a possibility of reducing the overall cost. However, a cost benefit analysis between answering the views directly from the base tables vs. materializing the views and answering some views from those materialized ones is required. For that purpose, let $C_b(V_{i,b})$ be the cost of answering a binned view $V_{i,b}$ from the base table. Then in Eq. 5, the cost of finding the top-1 binned view $(C(V_i))$, for the non-binned view V_i, can be rewritten as:

$$C(V_i) = \sum_{b=2}^{\frac{L}{w}} C_b(V_{i,b}) \qquad (7)$$

Notice $C(V_i)$ actually specifies the cost for option 1, where nothing is reused. For the other options, where reuse is involved, let $C_m(V_{i,b})$ be the cost of answering $V_{i,b}$ from a materialized view. Additionally, let the views be divided into two sets: (1) *Dependent Set*: the views that can be answered from $\mathbb{T}$ belong to the dependent set $\mathbb{P}$, and (2) *Independent Set*: the views that cannot be answered from $\mathbb{T}$ belong to the independent set $\mathbb{I}$. Particularly, the views in $\mathbb{I}$ need to

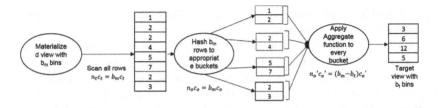

Fig. 3. Example of cost model for HashAggregate operator where $b_m = 8$ and $b = 4$

be answered from the base table. Let the cost of materializing a view $V_{i,b'}$ is $C_M(V_{i,b'})$, then Eq. 8 specifies the cost for option 3:

$$C(V_i) = \sum_{V_{i,b} \in \mathbb{P}} C_m(V_{i,b}) + \sum_{V_{i,b} \in \mathbb{T}} C_M(V_{i,b}) + \sum_{V_{i,b} \in \mathbb{I}} C_b(V_{i,b}) \tag{8}$$

Definition: Materialized View Selection for View Recommendation:
Given all the binned views $V_{i,b}$ for a non-binned view V_i, find a set $\mathbb{T}$ of views to materialize, which minimize the cost $C(V_i)$ of finding the top-1 binned view.

4 Methodology

Our proposed schemes in this section adapt and extend algorithms of materialized view selection towards efficiently solving the aggregate view recommendation problem.

4.1 mView: Greedy Approach

As explained int Sect. 2, the large number of possible binned views, makes the problem of finding the optimal binning for a certain view V_i highly challenging. An exhaustive brute force strategy is that given a certain non-binned view V_i, all of its binned views are generated and the utility of each of those views is evaluated. Consequently, the value of b that results in the highest utility is selected as the binning option for view V_i. However, this involves massive cost of processing all possible binned views.

In this work, we propose a novel technique *mView*, which instead of answering each query related to a view from scratch, reuses results from the already executed queries through view materialization. Particularly, *mView* maintains two sets of views; (1) $\mathbb{T}$: the views that are finalized to be materialized, (2) *Cand*: set of candidate views that can be added to $\mathbb{T}$ and consequently get materialized. The proposed technique *mView* adapts a greedy approach to determine $\mathbb{T}$ for materialization. Initially, *Cand* and $\mathbb{T}$ are empty. Then for a non-binned view V_i, a lattice as shown in Fig. 2 is constructed, using an adjacency list after identifying dependencies among the views. The search for the top-1 binned view starts from the binned view $V_{i,b}$ where $b = 1$. All of the views that are ancestors of $V_{i,b}$ in the lattice are added to the set *Cand*. Next, the benefit of materializing each view in *Cand* is computed.

We study in detail how to compute the benefit of materializing a view in the next paragraph. After benefit calculation, from the set *Cand*, a view V_m, which provides the maximum benefit is selected. V_m is added in $\mathbb{T}$ if it is not already in $\mathbb{T}$. Consequently, V_m is materialized and $V_{i,b}$ is generated from V_m. In next iteration *Cand* is set to empty again and the ancestors of the next binned view are added to *Cand*. This process goes on until all of the $V_{i,b}$ have been generated. Initial experiments show this approach reduced cost of Linear search by 25–30% Clearly, for this technique to work efficiently, a cost model is required to estimate the benefit of materializing views without actual materialization. Therefore, next we define that cost and benefit model.

Cost Benefit Analysis. As mentioned earlier, to decide which views are the best candidates for materialization, the cost and benefit of materialization needs to be analyzed. Specifically, we use processing time as our cost metric to measure performance of the schemes. In the linear cost model, the time to answer a query is taken to be equal to the space occupied by the underlying data from which the query is answered [1,7]. In this work, the same model is adopted with some modifications. Assume that the time to answer the aggregate query Q is related to two factors; (1) the number of tuples of the underlying view from which Q is answered, which is actually the number of bins of the ancestor view, and (2) the amount of aggregation required to answer Q. Normally, a relational DBMS uses HashAggregate as query execution plan for group-by queries. Particularly, in this study PostgreSQL is used as backend database, which uses HashAggregate as query execution plan for group-by queries. Hence, the cost model used by the query optimizer, particularly PostgreSQL consists of a vector of five parameters to predict the query execution time [14]; (1) Sequential page cost (c_s), (2) Random page cost (c_r), (3) CPU tuple cost (c_t), (4) CPU index tuple cost (c_i), and (5) CPU operator cost (c_o). The cost C_{HA} of the HashAggregate operator in a query plan is then computed by a linear combination of c_s, c_r, c_t, c_i, and c_o:

$$C_{HA} = n_s c_s + n_r c_r + n_t c_t + n_i c_i + n_o c_o$$

Where the values $n = (n_s, n_r, n_t, n_i, n_o)^{\mathbb{T}}$ represent the number of pages sequentially scanned, the number of pages randomly accessed, and so forth, during the execution.

Generally, for estimating cost of an operator, the values in vector n are estimated. However, in our case, the already known number of rows of a materialized view (i.e., number of bins of that view) and target view can be used for vector n. For instance, Fig. 3 shows the steps of the HashAggregate operator for generating a view with 4 bins from a view with 8 bins, and the cost incurred. Specifically, The operation of generating a view with b bins from a view with b_m bins has the following parameters:

- $n_s c_s$ & $n_r c_r$: c_s and c_r are the I/O costs to sequentially access a page and randomly access a page, while n_s and n_r are the number of sequentially and randomly accessed pages respectively. Generally, size of a page is 8 KB. Consequently, n_s depends on the page size, size of each row (let it be r), and the number of rows read, which is equal to the number of bins of the materialized view, i.e., $n_s = \frac{8\,KB}{r \times b_m}$. Furthermore, c_s and c_r depends on whether the data is fetched from the disk or it is already in cache. Particularly, this cost is negligible for the later case and that is the case in our model.

- $n_t c_t$: c_t is the cost of scanning each row and n_t is the number of rows scanned, which is equal to the number of bins of the materialized view, i.e., $n_t = b_m$.
- $n_i c_i$: c_i is the cost to place the row in a bucket (bin) using hashing and n_i is the number of rows hashed, which is equal to the number of bins of the materialized view, i.e., $n_i = b_m$.
- $n_o c_o$: c_o is the cost to perform aggregate operation such as sum, count etc., and n_o is the number of aggregate operations performed. If $V_{i,b}$ is answered from V_{i,b_m}, then there are b buckets and each bucket will require $\frac{b_m}{b} - 1$ aggregate operations, i.e., $n_o = b(\frac{b_m}{b} - 1) = b_m - b$.

Therefore, the cost of HashAggregate operator C_{HA} is:

$$C_{HA} = n_t c_t + n_i c_i + n_o c_o$$
$$C_{HA} = b_m c_t + b_m c_i + (b_m - b)c_o$$
$$C_{HA} = b_m(c_t + c_i + c_o) + b(-c_o)$$

The costs c_t, c_i, and c_o remain same for all queries. Therefore, we replace them with simple constants c and c' such that: $c = c_t + c_i + c_o$ and $c' = -c_o$. Hence,

$$C_{HA} = b_m \times c + b \times c'$$

Therefore, the cost of generating $V_{i,b}$ from materialized view V_{i,b_m} is:

$$C_m(v_{i,b}) = b_m \times c + b \times c' \tag{9}$$

Where c and c' are learnt through multi-variable linear regression. Consequently, the benefit of materializing a view V_{i,b_m} is computed by adding up the savings in the query processing cost for each dependent view $V_{i,b}$ over answering $V_{i,b}$ from the base table and subtracting the cost of materialization of V_{i,b_m}.

$$\mathbb{B}(V_{i,b_m}) = \sum_{V_{i,b} \in \mathbb{P}} [(C_b(V_{i,b}) - C_m(V_{i,b}))] - C_M(V_{i,b_m}) \tag{10}$$

In this section, we listed the details of our proposed technique $mView$ for the exhaustive search, which is also called $Linear$ search. When this scheme is applied to a non-binned view V_i, it results in a top-1 binned view, this is termed has horizontal search. Furthermore, applying this to every non-binned view, their corresponding top-1 binned views are identified and from there top-k views can be easily recommended, this is termed as vertical search. In our experiments, we differentiate between horizontal and vertical search and the scheme applied to each direction.

4.2 Materialized Views with MuVE

In [3,4], we argue that the deviation based utility metric falls short in completely capturing the requirements of numerical dimensions. Hence, a hybrid multi-objective utility function was introduced, which captures the impact of numerical dimension attributes in terms of generating visualizations that have: (1) interestingness $(D(V_{i,b}))$: measured using the deviation-based metric, (2) usability $(S(V_{i,b}))$: quantified via the relative bin width metric, and (3) accuracy $(A(V_{i,b}))$: measured in terms of Sum Squared Error (SSE). The proposed multi-objective utility function, was defined as follows:

$$U(V_{i,b}) = \alpha_D \times D(V_{i,b}) + \alpha_A \times A(V_{i,b}) + \alpha_S \times S(V_{i,b}) \qquad (11)$$

Parameters α_D, α_A and α_S specify the weights assigned to each objective, such that $\alpha_D + \alpha_A + \alpha_S = 1$. Furthermore, to efficiently navigate the prohibitively large search space $MuVE$ scheme was proposed, which used an incremental evaluation of the multi-objective utility function, where different objectives were computed progressively. In this section, we discuss how to achieve benefits of both the schemes, $mView$ and $MuVE$.

Selecting $\mathbb{T}$ while using $MuVE$ as search strategy is non-trivial, because of the trade-off between $MuVE$ and $mView$. In the $MuVE$ scheme, the benefit of cost savings comes from the pruning of many views and utility evaluations. A blind application of greedy view materialization, as in $mView$, may result in materialization of views that gets pruned because of the $MuVE$'s pruning scheme. The idea here is to estimate which views $MuVE$ will eliminate and exclude those views from the set of candidate views to materialize. To address this issue, we introduce a penalty metric, which is added to the benefit function. Therefore, a candidates view V_{i,b_m}, which has high certainty (represented as $CE(V_{i,b_m})$) of getting pruned by $MuVE$ gets a high reduction in its benefit of materialization. Particularly, a view gets pruned due to either of the two factors; (1) short circuit of deviation objective, the certainty of this pruning is represented as $CE_D(V_{i,b_m})$, and (2) early termination, certainty of getting early terminated is represented as $CE_E(V_{i,b_m})$. The certainty factor $CE(V_{i,b_m})$ is the sum of the certainty of pruning deviation evaluation $(CE_D(V_{i,b_m}))$ and certainty of getting early terminated $(CE_E(V_{i,b_m}))$.

$$CE(V_{i,b_m}) = CE_D(V_{i,b_m}) + CE_E(V_{i,b_m})$$

Therefore, the benefit of materializing a view in Eq. 10 is updated as:

$$\mathbb{B}(V_{i,b_m}) = \sum_{V_{i,b} \in \mathbb{P}} [C_b(V_{i,b}) - C_m(V_{i,b})] - [CE(V_{i,b_m}) \times C_M(V_{i,b_m})] \qquad (12)$$

The certainty of pruning deviation computation depends on the ratio of α_A and α_D. $MuVE$ uses a priority function to determine which objective to evaluate first, in other words $MuVE$ tries to prune the objective, which is not evaluated first. According to that function if α_A is greater than α_D there is a chance of pruning the deviation objective. We are interested in pruning deviation evaluation as it is the only objective that involves execution queries for target and comparison views.

$$CE_D(V_{i,b}) = \left\{ \begin{matrix} 0 \; for \; \frac{\alpha_A}{\alpha_D} < 1 \\ \frac{\alpha_A}{\alpha_D} \times 10 \; for \; \frac{\alpha_A}{\alpha_D} \geq 1 \end{matrix} \right\} \tag{13}$$

The certainty of early termination depends on α_S and b, higher value of α_S or b means the chance of getting early termination is high.

$$CE_E(V_{i,b}) = \left\{ \begin{matrix} 0 \; for \; \alpha_S < 0.5 \\ \alpha_S \times \frac{b}{L} \; for \; \alpha_S \geq 0.5 \end{matrix} \right\} \tag{14}$$

5   Experimental Evaluation

5.1   Experimental Testbed

We perform extensive experimental evaluation to measure the efficiency of top-k view recommendation strategies presented in this paper. Here, we present the different parameters and settings used in our experimental evaluation.

Setup: We built a platform for recommending visualizations, which extends the SeeDB codebase [13] to support view materialization based schemes presented in this paper. Our experiments are performed on a Corei7 machine with 16 GB of RAM. The platform is implemented in Java and PostgreSQL is used as the backend DBMS.

Schemes: We investigate the performance of the different combinations of the vertical and horizontal search strategies presented in [3] with *mView* proposed in this paper. Our naming convention for those combinations is represented as: *SearchH-SearchV*, where *SearchH* denotes the search strategy employed for horizontal search, whereas *SearchV* is the one for the vertical search. This leads to the following combinations: *Linear-Linear*, *MuVE-Linear*, and *MuVE-MuVE* as baseline schemes and *mView(Linear-Linear)*, *mView(MuVE-Linear)*, and *mView(MuVE-MuVE)* as proposed schemes.

Data Analysis: As in [13], we assume a data exploration setting in which a multi-dimensional dataset of diabetic patients[1] is analyzed. The DIAB dataset has 9 attributes and 768 tuples. The independent numeric attributes of the dataset are used as dimensions (e.g., age, BMI, etc.), whereas the observation attributes are used as measures (insulin level, glucose concentration, etc.). In our default setting, we select 3 dimensions, 3 measures, and 3 aggregate functions, which results in a maximum of 2961 possible views. In the analysis, all the α values are in the range $[0-1]$, where $\alpha_D + \alpha_A + \alpha_S = 1$. In the default setting, $\alpha_D = 0.2$, $\alpha_A = 0.2$, $\alpha_S = 0.6$, $k = 5$, and euclidean distance is used for measuring deviation, unless specified otherwise.

---

[1] https://archive.ics.uci.edu/ml/datasets/Pima+Indians+Diabetes.

Performance: We evaluate the efficiency and effectiveness of the different recommendations strategies in terms of two factors: (1) *Cost:* As mentioned in Sect. 3, the cost of a strategy is the total cost incurred in processing all the candidate binned views. We use wall clock time to measure the different components included in that cost namely, query execution time of target and comparison views, deviation computation time, and accuracy evaluation time. (2) *Relative Difference:* The ratio between cost of baseline schemes and the *mView* based schemes, i.e., $\frac{Cost\,of\,baseline - Cost\,of\,mView\,Scheme}{Cost\,of\,baseline\,scheme}$. Each setting is executed 10 times and then average is taken as the cost incurred.

5.2   Experiments

In the following experiments, we evaluate the performance of our technique *mView* under different parameter settings. As explained in Sect. 4 that *mView* scheme is used in combination with the baseline Linear scheme and optimized *MuVE* scheme. Additionally it was also mentioned that the blind materialization of views while using *MuVE* search strategy may not be the optimal solution. Therefore, for *mView(MuVE-MuVE)* and *mView(MuVE-Linear)* schemes an heuristic based method was proposed to predict the expected early termination and short circuit point. Figures 4 and 6 show the impact on cost, while Figs. 5 and 7 quantifies the percentage improvement achieved in terms of relative difference using the view materialization scheme.

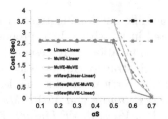

Fig. 4. Impact of α_A and α_S on cost, while $\alpha_D = 0.2$

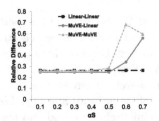

Fig. 5. Impact of α_A and α_S on relative difference, while $\alpha_D = 0.2$

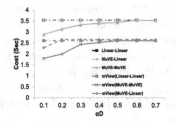

Fig. 6. Impact of α_A and α_D on cost, while $\alpha_S = 0.2$

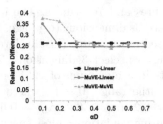

Fig. 7. Impact of α_A and α_D on relative difference, while $\alpha_S = 0.2$

In Figs. 4 and 5, α_D is set to constant 0.2 while α_A and α_S are changing. In particular, as shown in the figures, α_S is increased, while α_A is implicitly decreased and is easily computed as $\alpha_A = 1 - \alpha_D - \alpha_S$. Figure 4 shows that cost *mView(Linear-Linear)* is less than the baseline scheme *Linear-Linear*. This is because *mView* chooses such a set of views to materialize that saves aggregation time by generating them from the materialized views. Furthermore, Fig. 5 shows *mView(Linear-Linear)* reduces the cost by almost 30% as compared to the *Linear-Linear* scheme. Figure 4 also shows that using our proposed heuristic in *mView(MuVE-MuVE)* and the incremental view materialization of *mView*, the cost is further reduced. This is due to the reason that we avoided the unnecessary materialization of views which are eventually pruned by *mView(MuVE-MuVE)*. Furthermore, Fig. 5 shows *mView(MuVE-MuVE)* reduces the cost by almost 70% as compared to *MuVE-MuVE* at $\alpha_S = 0.6$.

In Figs. 6 and 7, α_S is set to constant 0.2 while α_A and α_D are changing. Figure 6 clearly shows that *mView* based three schemes have less cost compared to the other three schemes. The difference in cost for the *mView(MuVE-MuVE)* scheme is more than 30% at $\alpha_D = 0.1$ as shown in Fig. 7.

6   Conclusions

In this paper we presented a novel technique *mView* for recommending top-k binned aggregate data visualizations. The proposed scheme reuses the already executed views through materialization and answering the later queries from the materialized views. We defined a cost benefit model to decide which views can be reused later. We also proposed a heuristic based approach to predict the expected early termination and short circuit for *MuVE* based schemes. Our experimental results show that employing the *mView* technique for both *Linear* and *MuVE* based schemes offers significant reduction in terms of data processing costs.

References

1. Baralis, E., Paraboschi, S., Teniente, E.: Materialized views selection in a multidimensional database. In: VLDB, pp. 156–165 (1997)
2. Chaudhuri, S., Krishnamurthy, R., Potamianos, S., Shim, K.: Optimizing queries with materialized views. In: ICDE, pp. 190–200 (1995)
3. Ehsan, H., Sharaf, M.A., Chrysanthis, P.K.: MuVE: efficient multi-objective view recommendation for visual data exploration. In: ICDE, pp. 731–742 (2016)
4. Ehsan, H., Sharaf, M.A., Chrysanthis, P.K.: Efficient recommendation of aggregate data visualizations. IEEE Trans. Knowl. Data Eng. **30**(2), 263–277 (2018)
5. Gupta, H., Mumick, I.S.: Selection of views to materialize in a data warehouse. IEEE Trans. Knowl. Data Eng. **17**(1), 24–43 (2005)
6. Halevy, A.Y.: Answering queries using views: a survey. VLDB J. **10**(4), 270–294 (2001)
7. Harinarayan, V., Rajaraman, A., Ullman, J.D.: Implementing data cubes efficiently. In: SIGMOD, pp. 205–216 (1996)

8. Kandel, S., Parikh, R., Paepcke, A., Hellerstein, J.M., Heer, J.: Profiler: integrated statistical analysis and visualization for data quality assessment. In: AVI, pp. 547–554 (2012)
9. Key, A., Howe, B., Perry, D., Aragon, C.R.: VizDeck: self-organizing dashboards for visual analytics. In: SIGMOD, pp. 681–684 (2012)
10. Mafrur, R., Sharaf, M.A., Khan, H.A.: DiVE: diversifying view recommendation for visual data exploration. In: CIKM, pp. 1123–1132 (2018)
11. Srivastava, D., Dar, S., Jagadish, H.V., Levy, A.Y.: Answering queries with aggregation using views. In: VLDB, pp. 318–329 (1996)
12. Stolte, C., Tang, D., Hanrahan, P.: Polaris: a system for query, analysis, and visualization of multidimensional relational databases. IEEE Trans. Vis. Comput. Graph. 8(1), 52–65 (2002)
13. Vartak, M., Rahman, S., Madden, S., Parameswaran, A.G., Polyzotis, N.: SEEDB: efficient data-driven visualization recommendations to support visual analytics. PVLDB 8(13), 2182–2193 (2015)
14. Wu, W., Chi, Y., Zhu, S., Tatemura, J., Hacigümüs, H., Naughton, J.F.: Predicting query execution time: are optimizer cost models really unusable? In: ICDE, pp. 1081–1092 (2013)

Effective Community Search Over Location-Based Social Networks: Conceptual Framework with Preliminary Result

Ismail Alaqta[1,2(✉)], Junho Wang[1], and Mohammad Awrangjeb[1]

[1] School of Information and Communication Technology, Griffith University, Brisbane, Australia
ismail.alaqta@griffithuni.edu.au,
{j.wang,m.awrangjeb}@griffith.edu.au
[2] Department of Computer Science, Jazan University, Gizan, Saudi Arabia
ialaqta@jazanu.edu.sa

Abstract. Over the past decade, the volume of data has grown exponentially due to global internet service propagation. The number of individuals using the internet has expanded, especially with the use of social networks. Utilising GPS-enabled mobile devices, social networks have been labelled Location-based Social Networks (LBSN). This service enables users to share their current spatial information by 'checking-in' with their friends at different locations. This article proposes a conceptual framework to enhance the effectiveness of community search over LBSN. As users are more likely to look for people whom they share similar personalities and interests, these keywords plus the spatial information could help a lot in finding the most appropriate query-based social community. As a result, this paper aims to contribute to the existing body of knowledge as well as the industry in the field of community search (CS). In particular, this work is focusing on CS in the environment of LBSN to benefit from factors of spatial, keywords and time in order to enhance community search models by these factors. Therefore, in this study, we focus on the current state-of-the art of CS and the limitations of integrated models. The preliminary results confirm that user's checkins can present an alternative approach to produce and update the users' interests with which we use to boast effectiveness of attributed community search along with spatial information.

Keywords: Community search · User interests · Spatial graph

1 Introduction

Over the past decade, the volume of data has grown exponentially due to global internet service propagation. According to the latest report issued by the UN's international telecommunications union (ITU), the number of individuals using

© Springer Nature Switzerland AG 2019
L. Chang et al. (Eds.): ADC 2019, LNCS 11393, pp. 119–131, 2019.
https://doi.org/10.1007/978-3-030-12079-5_9

the internet exceeded 3.5 billion by 2017[1]. Social network applications, such as Facebook[2], Twitter[3] and Foursquare[4] are the most common internet applications. These applications have consequently attracted millions of users. For example, over 1.75 billion of Facebook users are active monthly[5]. Because they use GPS-enabled mobile devices, social networks have been called Location-based Social Networks (LBSNs). This service enables users to share their current spatial information by 'checking-in' with their friends at different locations. Foursquare, on which more than 30 million users are accommodated, receives millions of check-ins daily[6]. Other traditional social networks such as Facebook and Twitter[7] also provide users with the facility of check-ins, which can be utilised for many business purposes. In most cases, a check-in generates a triplet $\langle u, l, t \rangle$ indicating that user u checked-in at location l associated with spatial information $\langle x, y \rangle$ at a specific *time t*, which also shows that the user is temporally online. Consequently, this leads both industry and academia to consider the time dimension. People on social networks communicate with each other and this interaction is recorded with time. For example, consider social network users on the Gold Coast who are interested in a coffee shop at which their friends have already checked-in. This group of people have planned to meet up at a certain place and time. The coffee shop (e.g. Merlo)can also utilise its customers' profiles on Facebook to provide location-specific advertisements to potential customers, who might also be interested in other items offered by the coffee shop. However, this increases the complexity of the social network. Moreover, due to the vast development of online social networks, people can create and update their profiles. A huge amount of textual information is associated with users because they can express themselves easily through blogging. If a Flickr user utilises many keywords related to travelling (e.g. posts many photos about trips with keywords), these keywords help interested users to find people with similar interests. Basically, users are more likely to search for people with whom they share similar personalities and interests or those who share similar work and research areas. Users are progressively geo-coded and geo-positioned on social networks and there is increased availability of textual descriptions regarding interests, such as tourist attractions and cafes.

This research contributes to the existing body of knowledge as well as the industry in the field of community search. This work focuses on the social community in the environment of LBSN. Due to the variant data type of LBSN, the significance of this research can be classified into three dimensions: social relationships, attributes, and spatio-temporal. In terms of social matter, friend recommendations, in which the system searches for similar users to recommend

---

[1] http://www.itu.int/en/ITU-D/statistics.

[2] http://www.facebook.com.

[3] http://twitter.com.

[4] https://foursquare.com.

[5] http://www.statisticbrain.com/facebook-statistics.

[6] Foursquare statistics. https://foursquare.com/about/.

[7] www.twitter.com.

them to each other, is one of the most important outputs of community search. Moreover, as the users of LBSN can have keywords or tags to describe themselves or their businesses, a self-drive tour of a set of POIs or a minimum group of people, who share similar interests, could be achieved using an attributed community.

To model and search complex social graphs meaningfully, the simple graph model is often not adequate to capture many real-world social network datasets. As previously noticed from the examples, for most social networks, information is not only available about social connections but also about user demographics, preferences, actions performed, and so on. Combining both the explicit spatial association of a place and the implicit semantics of interaction with s place provides a unique opportunity for in-depth understanding of both places and users. Hence, in this research we investigate the possibilities of *spatio-attributed community search* to enrich the simple graph model.

2   Related Work

Community search is a community retrieval approach that aims to find a densely populated query-based on-line connected community (Fang et al. 2017; Li et al. 2015). For example, *k-core* (Seidman 1983) was utilised in (Li et al. 2015; Sozio and Gionis 2010). (Sozio and Gionis 2010) designed the first algorithm Global to retrieve the connected *k-core* that includes the vertex q. In detail, the problem was formulated as Q, a set of query nodes or seeds against a graph $G = (V, E)$ to retrieve a connected subgraph including Q. Thus, the authors suggested a function called the 'goodness function' f to measure the goodness of the subgraph. Moreover, this work (Sozio and Gionis 2010) considered subgraph density by using two other functions: the average and minimum degree of the subgraph nodes f_a and f_m, respectively.

2.1   Attributed Community Search

An attributed community is represented by vertices associated with text or keyword-named attributes. These attributes can effectively provide more features such as ease of interpretation and personalization (Fang et al. 2017). Recently, (Shang et al. 2017) proposed an attributed community search method, which was enhanced by (Huang et al. 2014), with a refining technique. The main idea was to reconstruct the graph based on topology-based and attribute-based similarities. The new reconstructed graph was called the TA-graph. Based on the TA-graph structure, an index named AttrTCP-index based on TCP-index (Huang et al. 2014) was created. Thus, queries that are on the new index AttrTCP-index return to communities that satisfy the queries. Moreover, (Fang et al. 2017) investigated the attributed community search by combining a cohesive structure and keyword. The data model in this study was similar to the previous one (Shang et al. 2017), specifically in keywords for which each vertex v is associated with a set of keywords. However, this work utilised the $k - core$ technique

(Seidman 1983) and the decomposition algorithm proposed in (Batagelj 2003) to find a cohesive structure called a connected $k - core$ denoted by $\widehat{k - core}$. More significantly, the study designed an index called the Core Label tree (CL-tree), which puts the $\widehat{k - core}$ and keywords in a tree structure. Based on the $k - core$ definition, the authors identify the research problem as given $G = (V, E)$, a positive integer k, a vertex $q \in V$ and a set of keywords $S \subseteq W$. In community search, index construction plays a key role due to the effective and efficient impact on results. Since cores can be nested (Batagelj 2003), the CL-tree index (Fang et al. 2017) was constructed. Obviously, a $\widehat{k - core}$ must contain $\widehat{(k + 1) - core}$. Thus, a tree structure is the most suitable data structure for such $k - cores$.

2.2 Spatial Community Search

Spatial graphs are on-line social networks on which users can share their location information, e.g. their position during check-ins. Spatial community search can perform community retrieval techniques, e.g. *k-core or k-truss* on a spatial social network. For example, given a Geo-Social Graph G, and a query vertex q, the task of spatial community search is to find a subgraph of G. This subsection reviews the most considerable works in terms of a spatio-social community search, as previously reviewed works assume non-spatial graphs (Cui et al. 2013; 2014; Huang et al. 2014; Li et al. 2015; Sozio and Gionis 2010). It can be said that a recent work named *spatial-aware community* (SAC) (Fang et al. 2016) has adopted the concept of minimum degree, which basically depends on the $k - core$ technique. SAC is a subgraph denoted by $H = (V_H, E_H)$, which needs to satisfy the following:

- Connectivity, $G_q \in G$ is connected and q exists.
- Structure cohesiveness $\Rightarrow$ all vertices are intensively linked in H.
- Spatial cohesiveness $\Rightarrow$ all vertices are almost at the same spatial location.

Compared to traditional CS works, condition three is intuitively what distinguishes SAC. So, spatial cohesiveness in SAC is defined to achieve a minimum covering circle (MCC) with the smallest radius. The formal definition is that given a set of vertices $\mathbf{S}$, the MCC of $\mathbf{S}$ is the spatial circle that contains all vertices in S with the smallest radius. SAC follows the two-step framework: (1) find a community S of vertices, based on some CS algorithm (Sozio and Gionis 2010); and (2) find a subset of S that satisfies both structure and spatial cohesiveness.

All reviewed methods consider social constraints. Some reinforce social queries with extra constraints, e.g. keywords, location, and time. However, there is a lack of integration of all constraints into one CS framework. Therefore, this article proposes a conceptual framework to enhance the effectiveness of community search models over LBSN. The enhancement has been enforced by integrating compatibly text-mining techniques with a community search model as demonstrated in Fig. 1.

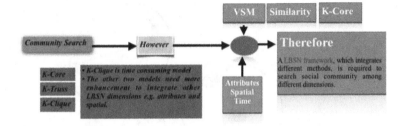

Fig. 1. Research gap

3 Methodology

This research has developed a hybrid approach, which aims to search for desired query-based social communities over LBSN. Our hybrid approach considers three different dimensions, *including keywords, location, and time*, which significantly enhance the effectiveness of social community search outputs. Therefore, the approach has combined different methods in which each method is required to achieve its research objectives under one framework.

3.1 Problem Formulation

In this section, we provide definitions that will be used throughout the paper. Also, this section provides the problem statement followed by an example to elaborate our research problem.

Data Model: We consider the location-based social network $G = (V, E, X)$ as an attributed graph, where V is a set of all users. Each edge $e(u, v) \in E$ indicates that a friendship exists between two users. X denotes a matrix $[X]\, n \times l$ where l is the number of all possible distinct keywords W, which are associated with places P that have been visited by users in form of 4-tuple check-in point CK. So, $CK = \{\langle u_i, p_k, t, W_{p_k} \rangle | u_i \in V, p_k \in P\}$ where p_k is identified by a unique GPS coordinate and t is a time-stamp when a user u_i checked-in p_k. For example, in Fig. 2 there are nine users, i.e. $u_1, ..., u_9$. Some conform with the conditions of inducing dense subgraphs. For instance, $\langle u_1, u_2, u_3, u_4 \rangle$, $\langle u_7, u_8, u_9 \rangle$ are two subsets, which form socially dense subgraphs. Moreover, our example shows that users could visit places either as a group or individually, e.g. $\langle u_1, u_2, u_3, u_4 \rangle$ checked-in at the same time t_1 and the same place as well, which results in keeping a dense spatio-temporal relationship. Later, we will learn how to define our query model to retrieve communities. Based on our data model, we give the following definitions followed by the query model.

Query Model: The main goal of our framework is to search the community of a location-based social graph. Our query model is maintained by several constraints that need to be satisfied to return an Attri-Spatial Social Community.

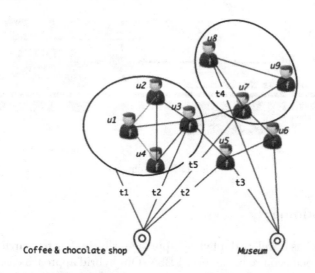

Fig. 2. Motivation example of location-based social network

Let us consider an example of a LBSN user u, who enquires where her friend *Majid* and his friends went for *coffee and chocolate* last year. People acquire each other's choices and interests e.g. user u likes the choice of her friend, *Majid*, in coffee and chocolate. To model this query, let q be a query that needs to retrieve all u's friends who visited a place p_u in *time t*. In addition, the query q has a keyword constraint that has the keywords of *coffee and chocolate*. Similarly, q can be asked to retrieve all places P_u visited by u's friends in a certain *time t* and keywords. The result set of q is $V' \subset P_u \subset V$.

3.2 Definitions

Based on the data and query model in the previous section, the following are definitions with which to draw up our framework.

Definition 1 (*User's Interests*): Let $U \subseteq V$ be a set of Users U who have CKs. Each user $u \in U$ is associated with a vector of keywords W_u. These keywords are extracted from places P, that have been visited by the user u, to represent users' interests as vectors in the space model.

Definition 2 (*Interest Weight*): Let each interest be $w \in W$ where W is a keyword set. Each $w \in u$ is associated with a weight named Relevance Score RS to indicate the interests' weights $\forall u \in U$.

Definition 3 (*Similarity-Based Graph SBG*): Given an attributed graph $G = (V, E, W)$, the SBG is a refined social graph constructed by computing interest-based similarities, which can be measured using a similarity function. The SBG can enhance the relationships between users regarding the user's interests. In addition, the SBG helps in returning accurate, query-based communities.

Definition 4 (*k-core* (Seidman 1983)): Given an integer $k \geq 0$, an existing connected subgraph $G(V')$ is called $\widehat{k - core}$ iff $\forall v \in G(V')$ has $deg(v) \geq k$, and $G(V')$ is connected.

Definition 5 (*Core Relevance Score CRS*): Given a query-based attributed subgraph $H \subseteq G$, and an interest w, the weight of interest w is RS as in Definition 2. Thus, CRS computes the relation between each subgraph H_i and their interest weight $\forall w \in H$

$$CRS_{H_i} = \frac{\sum_{w \in H_i} RS}{|H_i|} \tag{1}$$

where $|H_i|$ is the number of users $u \in H_i$.

Problem Definition: Given an undirected LBSN $G = (V, E)$, an integer k, $q \in V$, $w \in W_q$ and r, returns a subgraph $G_q \subset G$, which satisfies the following properties (Table 1):

Table 1. Query property

$Q = (q, k, w, r)$	
Property	Meaning
Connectivity	$G_q \subseteq G$ is connected and $q \in G_q$
Structure cohesiveness	$\forall v \in G_q, deg(v) \geq k$
Interest cohesiveness	Ensures that G_q has maximum $\widehat{CRS}$
Spatial range	A given radius r that ensures $\widehat{k - cores}$ are located within the range

3.3   Framework

In this section, we explain the framework and demonstrate how the three phases can interact with each other as one architecture. As shown in Fig. 3, our proposed architecture initially processes tags associated with places visited by users, followed by processing the social graph by linking each user with a vector of interests; each interest also carries weight. Once the attributed graph is created with associated vectors, we select all the pairs, which are unacquainted, but have at least one common neighbour, to measure the similarity between each pair. Based on comparing the similarity with a given threshold θ, we add any pairs that satisfy the minimum θ. Next, we compute the core decomposition. Finally, an index named AttriSpatial is created. The index is composed of two components - keywords and spatial -to handle the query $Q = (q, k, w, r)$.

User-Keyword TF-IDF Matrix. In this model, the well-known information retrieval model *Term Frequency Inverse Document Frequency TF-IDF* has been adapted to calculate the weight of users' interests. Accordingly, keywords W_p, extracted from places P_u that were checked into by a user u, are regarded as terms

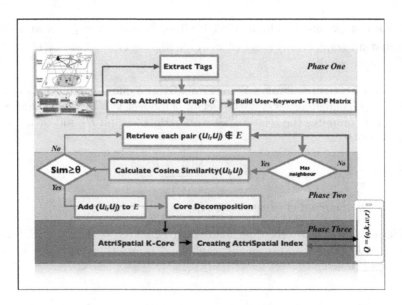

Fig. 3. Conceptual framework

and each user u is regarded as a document. Thus, each keyword of interest $w \in$ *user u* is represented as a dimension in the vector space model and, consequently, each user u_i is represented as a vector. To compute the weight of each keyword RS defined by Definition 3, the following is the *TFIDF* model.

$$RS_{i,j} = \frac{f_{w_i,u_j}}{\sum_{w_i \in u_j} f_{w_i,u_j}} \cdot log \frac{|U|}{|u \in U : w_i \in u|} \tag{2}$$

where $f_{w,u}$ is the keyword frequency for each *user u*, and $\sum_{w \in u} f_{w,u}$ is the total number of a user's keywords W_u, which were acquired via place check-ins P_u and $log \frac{|U|}{|u \in U : w \in u|}$ calculates the inverse user frequency of keywords w.

Intra-similarity of Users. After phase one is complete, phase one output, which is the result of Algorithm 1, is used to input phase two. In this phase, the intra-similarity of users depends on each keyword relevance score associated with them. In such a case, we adopt the cosine similarity to calculate the similarity between a pair of users. Each user is represented by an interest weight vector x_i.

$$attr - sim(v_i, v_j) = \frac{x_i \cdot x_j}{\|x_i\|_2 \cdot \|x_j\|_2} \tag{3}$$

To guarantee that there is a minimum familiarity, each pair must have at least one common neighbour friend before adding them as friends.

On-line Query Processing. The aim of phase three is to search for the best cohesive query-dependent communities based on the attributed-spatial constraints. The cohesiveness of communities, returned by queries, attempts to align all constraints: keywords, social, and spatial. As a result of the output of the previous phases, during phase three we have produced a technique named *AttriSpatial K-Core*. In addition, a ranking function, called *Core Relevance Score CRS*, can be derived from the technique. The task of the derived function is to rank the retrieved communities based on the interest weight from phase one as well as the community structure from phase two. More significantly, this technique is employed to construct an efficient hybrid index to improve our dimensional query processing.

Data: $G = (V, E)$, *Users Check-ins* $CK = \{\langle u_i, p_k, t, W_{p_k}\rangle\}$
Result: *Attributed Graph with weighted attributes* $G = (V, E, X)$
begin
    initialisation
    $V \longleftarrow U$
    $W \longleftarrow \emptyset$
    $X[V, W]$
    for $v \in V$ **do**
      | $X[v, RetKeyword(v)]$
    end
    for $v, w \in X[]$ **do**
      | *Compute RS*
    end
end

Algorithm 1. Attributed graph extracting

4 Case Study of a Location-Based Social Network

For the sake of producing preliminary results, a conceptual framework (Armenatzoglou et al. 2013) has been reimplemented with modifications. Also, a dataset called *Weeplaces* has been used to evaluate our proposed hybrid approach.

4.1 Dataset

Weeplaces is a dataset (Liu et al. 2014) that has been collected from a website named Weeplaces, in which users' check-in activities can be visualised in LBSN. It has been integrated using the API of other well-known LBSNs, e.g. Facebook Places, Foursquare, and Gowalla. The dataset contains more than 7.5 million check-ins by 15,799 users across 971,309 geolocations. Most importantly, we have revealed that Weeplaces has a reasonable number of communities. As demonstrated in Fig. 4, various values of k affect the order of cores as defined in Definition 4.

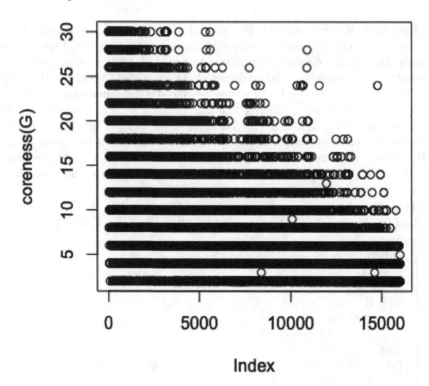

Fig. 4. Coreness distribution

4.2 Preliminary Results and Discussion

Setup. Two different data storage approaches have been employed for social and spatial layers. The two storage schemes have been implemented on MongoDB, a document-oriented database with the Python programming language. At the social layer, the social graph has been stored as a set of documents regarding the adjacency list representation. Moreover, each user has a list of keywords representing weighted interests. The spatial layer, which shows each place visited by a user, creates a document to represent that place. Thus, this document has a place ID, the visiting users' IDs, the tags that describe the place, and location coordinates.

Results. It is worth noting that our dataset has keywords to describe places that have been checked into by users. Keywords help us to understand users' interests by investigating the keyword frequency of each user compared to other users. Initially, we must represent the keywords distribution for the entire dataset over 100 users, in which each user is represented as a document. As shown in Fig. 5, on axis x, we plotted the graph attributes extracted by Algorithm 1. These attributes were associated with the weights calculated by the TFIDF schema in

AVG Interest TFIDF-based Weghit

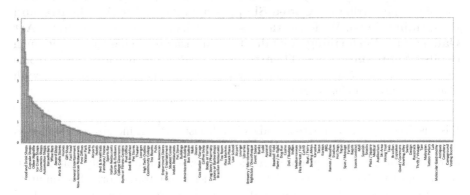

Fig. 5. Interest weights

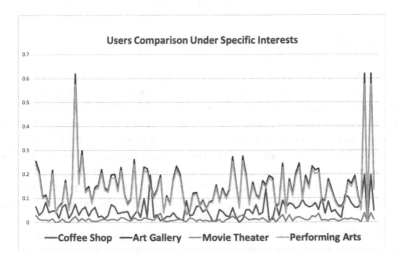

Fig. 6. Interest user comparison

Sect. 3.3. The goal is to differentiate between users using their interests. TFIDF is usually utilised as features to represent users as documents.

In Fig. 5, we grouped interests based on the global average TFIDF. The average will help us to select the appropriate threshold to retrieve query-based communities from which members contain attributes that are associated with greater weights than the threshold. The distribution in Fig. 5 demonstrates that a certain number of interests (graph attributes) are associated with weight score. This places these attributes in the area under the curve at which users share significant and representative interests, according to the feature selection technique, TFIDF. This leads us to the main contribution of our research; there may be other similar users who could increase the possibility of retrieving cohesive communities, in terms of interests.

We looked closely at Fig. 6, in which users are compared to each other under specific graph attributes (**Coffee Shop, Art Gallery, Movie Theatre and Performing Arts**). We found that users were had very similar interests (**Art Gallery and Performing Art**) due to their visits to similar places, although (**Coffee Shop and Movie Theatre**) were lower. Consequently, this outcome encouraged us to continue semantically investigating keywords associated with places that users prefer to visit.

Discussion. Previous related studies emphasise the significance of attributed graphs in community search. However, most of these studies investigated attributes, such as users' interests, as static keywords in community search. This study aims to integrate text analytic techniques into a framework of community search to keep users' profiles updated with their interests. This guarantees that communities retrieved by the framework are larger, more familiar, and more accurate than communities returned by community search models only.

5   Conclusion and Future Work

In this article, a recent body of knowledge regarding community search over a social graph has been reviewed. Specifically, two dimensions, including attributes and geolocations, have been investigated in more detail. This paper proposes a conceptual framework with preliminary results. Technically, this paper has shown that we can enrich users' profile interests by extracting keywords associated with places they visit. These interests have been analysed to produce an attributed social graph. As this work is part of ongoing research, in the future, we will conduct extensive experiments on various datasets of LBSN. Experimental work will include updating the social graph based on interest-based similarity among users as indicated in phase two. We will then create an efficient hybrid index that can handle different types of data, as explained in phase three. Furthermore, one of our future goals is to perform several comparisons, either using a baseline system or state-of-the-art work. Our future work will include the validation and feasibility of the proposed framework, in terms of effectiveness and efficiency.

References

Armenatzoglou, N., Papadopoulos, S., Papadias, D.: A general framework for geo-social query processing. Proc. VLDB Endowment **6**(10), 913–924 (2013). ISSN 21508097

Batagelj, V.: Efficient Algorithms for Citation Network Analysis. Networks, pp. 1–27 (2003)

Cui, W., Xiao, Y., Wang, H., Lu, Y., Wang, W.: Online search of overlapping communities. In: Proceedings of the 2013 International Conference on Management of Data - SIGMOD 2013, p. 277 (2013). ISSN 07308078

Cui, W., Xiao, Y., Wang, H., Wang, W.: Local search of communities in large graphs. In: Proceedings of the 2014 ACM SIGMOD International Conference on Management of Data - SIGMOD 2014, vol. 1, pp. 991–1002 (2014). ISSN 07308078

Fang, Y., Cheng, R., Li, X., Luo, S., Hu, J.: Effective community search over large spatial graphs. Proc. VLDB Endowment 9(12), 1233–1244 (2016). ISSN 21508097

Fang, Y., Cheng, R., Chen, Y., Luo, S., Hu, J.: Effective and efficient attributed community search. VLDB J. 26(6), 803–828 (2017). ISSN 0949877X

Huang, X., Cheng, H., Qin, L., Tian, W., Yu, J.X. Querying k-truss community in large and dynamic graphs. In: Proceedings of the 2014 ACM SIGMOD International Conference on Management of Data - SIGMOD 2014, vol. 2, pp. 1311–1322 (2014). ISSN 07308078

Li, R.-H., Qin, L., Yu, J.X., Mao, R.: Influential community search in large networks. Proc. VLDB Endowment 8(5), 509–520 (2015). ISSN 2150-8097

Liu, Y., Wei, W., Sun, A., Miao, C.: Exploiting geographical neighborhood characteristics for location recommendation. In: Proceedings of the 23rd ACM International Conference on Information and Knowledge Management, CIKM 2014, pp. 739–748. ACM, New York (2014). ISBN 978-1-4503-2598-1. https://doi.org/10.1145/2661829.2662002

Seidman, S.B.: Network structure and minimum degree. Soc. Netw. 5(3), 269–287 (1983). ISSN 03788733

Shang, J., Wang, C., Wang, C., Guo, G., Qian, J.: An attribute-based community search method with graph refining. J. Supercomput. 1(1), 1–28 (2017). ISSN 1573-0484

Sozio, M., Gionis, A.: The community-search problem and how to plan a successful cocktail party. In: Proceedings of the ACM SIGKDD International Conference on Knowledge Discovery and Data Mining, pp. 939–948 (2010)

Demo Paper

Context-Aware Visualization of Entity-Entity Relationships in a Document Corpus

Andreas Schmidt[1,2(✉)], Philipp Kief[1], and Steffen Scholz[2]

[1] University of Applied Sciences, Karlsruhe, Germany
philipp.kief@gmx.de
[2] Karlsruhe Institute of Technology, Karlsruhe, Germany
{andreas.schmidt, steffen.scholz}@kit.edu

Abstract. The paper presents an interactive graphical environment, which enables the detection and graphical visualization of concepts in a document or a document collection. Concepts are expressed by (multiple) entities extracted from the document and by the relationships between them. The tool offers an entity-centered view, which graphically shows the most important relationships of a central entity or entity-group, consisting of multiple co-occurring entities. By specifying prefixes and an additional available type system, complex filters can be created that allow the disclosure of various relationships between entities. Entities and their relationships are determined at the time of indexing and stored in appropriate data structures, so that an interactive search and exploration of relationships between entities is easily possible at runtime. The tool is available for online demonstration at https://www.smiffy.de/CoOcViz (credentials: adc2019, password: demo).

Keywords: Named entities · Document semantic · Entity co-occurrence · Visualization · Navigation

1 Introduction

In recent years there has been a series of advances in Natural Language Processing research and with technologies such as Named Entity Recognition (NER) [1] and Named Entity Disambiguation (NED) [2] we now have tools which can support us in analyzing the content and sentiment of unknown texts. So, for the understanding of a news article it is extremely helpful to know which entities, such as persons, places, organizations, times appear in it. By examining these entities to determine where they appear within a document and together with which entities, first important conclusions can be drawn about an article or even an article collection. Shifting the focus from the document to the entity, it is also very interesting to find out how a given entity relates to other entities in a document corpus. Transferred to the medical field, for example, it is certainly valuable to know that the combination of two specific drugs is repeatedly mentioned with unwanted skin irritation.

© Springer Nature Switzerland AG 2019
L. Chang et al. (Eds.): ADC 2019, LNCS 11393, pp. 135–138, 2019.
https://doi.org/10.1007/978-3-030-12079-5_10

2 Demonstrator

The functionality of the visualization tool is as follows. At the beginning, an entity is selected for which the relationship network is to be analyzed. The selection is made by specifying the entity using prefixes. An auto-suggestion and completion service offers suitable entities according to the prefix(s) entered. After selecting the desired entity, a view as shown in Fig. 1(a) is displayed. The chosen entity is located in the center (red). The most n-relevant entities related to it are grouped around it. The thickness of the edge expresses the degree of strength of the relationship. Each displayed entity can be selected as a new central entity by mouse click. Alternatively you can click on an edge. In this case an entity-group is created, consisting of the entities related by the edge (<Olympic Games>, <Sochi>). The result of such a click is shown in Fig. 1(b). The entity-group forms the new center and around it are shown entities that are related to this entity-group. What you can also see are individual dashed edges between the blue entities. This means that there is also a strong co-occurrence between these entities. These edges are also clickable and create a new entity group as new central node.

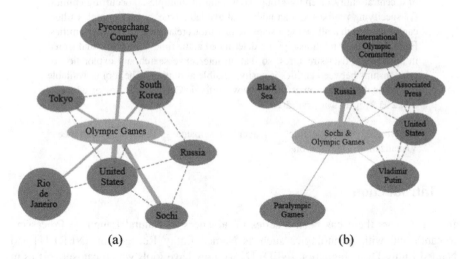

(a)   (b)

Fig. 1. Screenshots of visualization tool with *<Olympic Games>* resp. *<Olympic Games>* & *<Sochi>* as central nodes (Color figure online)

The tool also offers a series of filter options. By specifying a prefix, only the entities that fulfill a certain prefix are displayed. Figure 2(a) shows this using the prefix 'summer'. Already during the input of the prefix, a two-column selection box is displayed, which shows all qualifying entities in the left column (here: all summer olympiads). Selecting one of these entities creates an entity-group with the previous central entity (or group) and the newly selected entity as new central node. On the right side of the entities, categories are displayed which also pass the prefix match and which contain entities that are related to the current central entity. For example, 62 entities related to the entity *<Olympic Games>* belong to the category *<Swimmers at the 2012*

Summer Olympics>. Selecting a category causes a filter to be set that only displays entities in that category. Figure 2(b) shows a category filter (*<Medalists 2014 Winter Olympics>*). In this case only athletes who have won a medal at the 2014 winter games will be shown in the graph. In Fig. 2(b) you can additionally see that the number of displayed nodes in the graph can be specified. The number of entities related to the central entity (red) can be varied between 5 and 50. In addition, the m (between 0 and 10) most important entities to these entities can also be displayed. This is particularly helpful if the related entities have further entities in common or if entities are not very familiar but can be classified by the additional entities (e.g. country of an athlete). Another important feature is the time aspect. As can also be seen in Fig. 2(b), it can be specified which news articles (restricted by year) will be used for the calculation of the relatedness measure. This can make important changes in time visible.

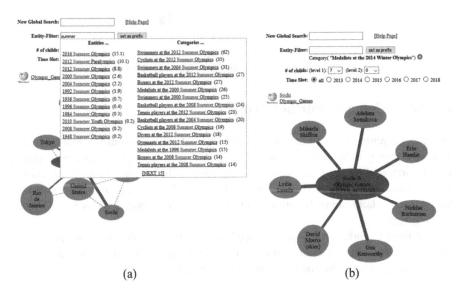

(a)   (b)

Fig. 2. Auto-completion for a given prefix (a) and an applied category filter *<Medalists at the 2014 Winter Olympics>* (b) (Color figure online)

3 Related Work

Schmidt et al. [3] have published a prior work on this topic, where related entities matching a prefix are suggested in the search interface to speed up query formulation for the entity-based search engine STICS [4]. The main finding was that context-sensitive suggestions have to be made. Figure 3 gives an example for this behavior. Depending on the context (none, *<Donald Trump>*, *<Hollywood>*) different suggestions were made for the same prefix. This work, on the contrary, explicitly shows relations between entities in a graphical and navigational manner. Indeed, the data structures used are partly the same. ESPRESSO [5] considers the relationships between two entity sets. They identify dense subgraphs with strong relationships to both sets. In contrast to our work, they focus on entity-sets and not on document sets.

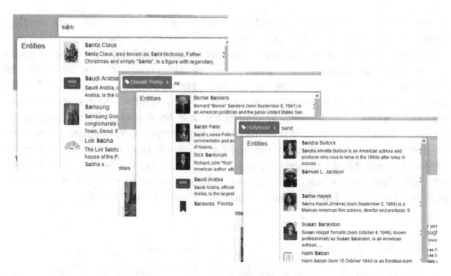

Fig. 3. STICS, context sensitive autocompletion & suggestion service

4 Conclusion and Further Work

The paper reported on a system for identifying and analyzing entities and their relationships among a document corpus. The system uses a graph based representation which further allows the navigation along entity relationships as well as filtering relationships based on prefixes and/or categories. The system not only considers bidirectional relationships, but also relationships between more entities (the so-called entity groups).

For further work, we intend to extend the interface, so that the documents most relevant to an entity or entity group can be inspected along with the parts of the documents, providing the most needed boost for that entity or entity type.

References

1. Nadeau, D., Sekine, S.: A survey of named entity recognition and classification. Linguist. Investig. **30**(1), 3–26 (2007)
2. Hoffart, J.: Discovering and disambiguating named entities in text, Ph.D. dissertation. Universität des Saarlandes, Saarbrücken (2015)
3. Schmidt, A., Hoffart, J., Milchevski, D., Weikum, G.: Context-sensitive auto-completion for searching with entities and categories. In: Proceedings of the 39th International ACM SIGIR Conference, pp. 1097–1100 (2016)
4. Hoffart, J., Milchevski, D., Weikum, G.: STICS: searching with strings, things, and cats. In: Proceedings of the 37th International ACM SIGIR Conference, pp. 1247–1248 (2014)
5. Seufert, S., Berberich, K., Bedathur, S.J., Kondreddi S.K., Ernst, P., Weikum, G.: Espresso: explaining relationships between entity sets. In: Proceedings of the 25th ACM International CIKM Conference, pp. 1311–1320 (2016)

Author Index

Printed in the United States
By Bookmasters